MOLECULAR BIOLOGY
INTELLIGENCE
UNIT

Protein Movement Across Membranes

Jerry Eichler, Ph.D.
Department of Life Sciences
Ben Gurion University
Beersheva, Israel

LANDES BIOSCIENCE / EUREKAH.COM
GEORGETOWN, TEXAS
U.S.A.

SPRINGER SCIENCE+BUSINESS MEDIA
NEW YORK, NEW YORK
U.S.A.

Protein Movement Across Membranes

Molecular Biology Intelligence Unit

Landes Bioscience / Eurekah.com
Springer Science+Business Media, Inc.

ISBN: 0-387-25758-6 Printed on acid-free paper.

Springer Science+Business Media, Inc., 233 Spring Street, New York, New York 10013, U.S.A.
http://www.springeronline.com

Please address all inquiries to the Publishers:
Landes Bioscience / Eurekah.com, 810 South Church Street, Georgetown, Texas 78626, U.S.A.
Phone: 512/ 863 7762; FAX: 512/ 863 0081
http://www.eurekah.com
http://www.landesbioscience.com

Printed in the United States of America.

9 8 7 6 5 4 3 2 1

Library of Congress Cataloging-in-Publication Data

Protein movement across membranes / [edited by] Jerry Eichler.
p. ; cm. -- (Molecular biology intelligence unit)
Includes bibliographical references and index.
ISBN 0-387-25758-6 (alk. paper)
1. Proteins--Physiological transport. 2. Cell membranes. I. Eichler, Jerry. II. Series: Molecular biology intelligence unit (Unnumbered)
[DNLM: 1. Protein Transport--physiology. 2. Membrane Transport Proteins--physiology. 3. Membranes. QU 120 P967 2005]
QP551.P69723 2005
572'.696--dc22

2005016424

CONTENTS

EDITOR

Jerry Eichler
Department of Life Sciences
Ben Gurion University
Beersheva, Israel
Chapter 3

CONTRIBUTORS

Ben C. Berks
Department of Biochemistry
University of Oxford
Oxford, U.K.
Chapter 6

Kenneth K.Y. Chan
Department of Biochemistry
and Molecular Biology
University of British Columbia
Vancouver, Canada
Chapter 2

Ian Collinson
Department of Biochemistry
School of Medical Sciences
University of Bristol
Bristol, U.K.
Chapter 4

Ross E. Dalbey
Department of Chemistry
The Ohio State University
Columbus, Ohio, U.S.A.
Chapter 5

Franck Duong
Department of Biochemistry
and Molecular Biology
University of British Columbia
Vancouver, Canada
Chapter 2

Ralf Erdmann
Institut für Physiologische Chemie
Abteilung für Systembiochemie
Ruhr-Universität Bochum
Bochum, Germany
Chapter 10

Ramanujan S. Hegde
Cell Biology and Metabolism Branch
NICHD
National Institutes of Health
Bethesda, Maryland, U.S.A.
Chapter 1

J. Michael Lord
Department of Biological Sciences
University of Warwick
Coventry, U.K.
Chapter 7

Antoine P. Maillard
Department of Biochemistry
and Molecular Biology
University of British Columbia
Vancouver, Canada
Chapter 2

Mark Paetzel
Department of Molecular Biology
and Biochemistry
Simon Fraser University
Burnaby, Canada
Chapter 5

Tracy Palmer
Department of Molecular Microbiology
John Innes Centre
Norwich, U.K.
Chapter 6

Doron Rapaport
Institut für Physiologische Chemie
der Universität München
Munich, Germany
Chapter 9

Lynne M. Roberts
Department of Biological Sciences
University of Warwick
Coventry, U.K.
Chapter 7

Frank Sargent
School of Biological Sciences
University of East Anglia
Norwich, U.K.
Chapter 6

Danny J. Schnell
Department of Biochemistry
and Molecular Biology
University of Massachusetts
Amherst, Massachusetts, U.S.A.
Chapter 8

Matthew D. Smith
Department of Biochemistry
and Molecular Biology
University of Massachusetts
Amherst, Massachusetts, U.S.A.
Chapter 8

Sven Thoms
Institut für Physiologische Chemie
Abteilung für Systembiochemie
Ruhr-Universität Bochum
Bochum, Germany
Chapter 10

PREFACE

For cells to function properly, correct protein localization is essential. This is true for both prokaryotes, i.e., Bacteria and Archaea, where proteins may be directed outside the confines of the cytoplasm to take up residence in the plasma membrane or beyond, as well as for eukaryotes, which also have to ensure that selected proteins are correctly distributed between the various organelles found inside the cell. Such non-cytoplasmic proteins must, therefore, be effectively recognized and targeted to their designated subcellular locations, where translocation across one or more membranes takes place. Across evolution, cells have developed complex systems dedicated to the transfer of proteins across a variety of biological membranes. In this volume, aimed at both the newcomer seeking an introduction to the subject and the expert wanting to keep abreast of recent discoveries in the field, the reader will learn about various aspects of protein translocation across a variety of membranes.

Translocation of exported proteins in each of the three domains of Life is the focus of the first four chapters. In Chapter 1, recent findings and outstanding questions regarding protein translocation across the membrane of the endoplasmic reticulum, the first step on the eukaryal secretory pathway, are presented. Chapter 2 provides insight into the latest discoveries in bacterial Sec-dependent translocation. In Chapter 3, current understanding of protein translocation in Archaea is discussed. Chapter 4 reveals how structural biology joins genetics and biochemistry as experimental approaches being employed to better understand translocation through the Sec translocon.

Indeed, as we learn more about protein translocation, previously hidden aspects of the process are being uncovered. Chapter 5 addresses strategies adopted by Bacteria for the integration of membrane proteins from a structural perspective. In Chapter 6, the twin arginine transport system, a more recently-defined translocation system largely employed for the transit of folded and complexed proteins across the membrane, is discussed. Chapter 7 describes how the endoplasmic reticulum exploits the Sec-based translocon for retrograde translocation of defective proteins back into the cytosol, where they undergo proteasome-based degradation.

Finally, several chapters examine the manner by which proteins are imported into different cellular organelles. Playing central roles in cellular metabolism, the chloroplast, mitochondria and peroxisome obtain most, if not all, of their proteins from sites of synthesis in the cytoplasm. Chapter 8 addresses how protein translocation into and across the membranes surrounding the chloroplast and the various sub-compartments contained therein takes place. Chapter 9 considers how proteins are delivered from outside the mitochondria into either the matrix or the inter-membrane space, as well as how outer and inner membrane proteins are inserted. In Chapter 10, current understanding of one of the least-well described protein import systems, namely that of the peroxisome, is considered.

With biological investigators now able to simultaneously address numerous complex processes at the cellular, system and even entire organism levels, a more thorough understanding of protein translocation is essential. This volume represents a step in that direction.

Jerry Eichler
Department of Life Sciences
Ben Gurion University
Beersheva, Israel

Chapter 1

Protein Translocation Across the Endoplasmic Reticulum Membrane

Ramanujan S. Hegde*

Abstract

Proteins to be secreted from eukaryotic cells are delivered to the extracellular space after trafficking through a secretory pathway composed of several complex intracellular compartments. Secretory proteins are first translocated from the cytosol into the endoplasmic reticulum (ER), after which they travel by vesicular trafficking via various intermediate destinations en route to the plasma membrane where they are released from the cell by exocytosis. By sharp contrast, secretion in prokaryotes involves the translocation of proteins directly across the plasma membrane. While these two systems are superficially dissimilar, they are evolutionarily and mechanistically related. This relationship between the prokaryotic and eukaryotic systems of secretion forms the backdrop for this chapter focused on protein translocation into the ER. In the first part of this chapter, the essential steps and core machinery of ER translocation are discussed relative to evolutionarily conserved principles of protein secretion. The last section then explores the concept of regulation, a poorly understood facet of translocation that is argued to be evolutionarily divergent, relatively specific to the ER, and likely to be most highly developed in metazoans.

Reductionistic View of ER Translocation

The eukaryotic secretory pathway is thought to have evolved by a series of steps that were initiated by specialization of the prokaryotic plasma membrane (Fig. 1). This specialized region of membrane was then expanded, internalized, and eventually subdivided into many compartments. Hence, the lumenal space of compartments in the secretory pathway is topologically equivalent to the extracellular space, and the transport of proteins across the prokaryotic plasma membrane is directly analogous to transport into the ER. Both processes face the same basic challenges: (a) substrates to be transported need to be **recognized**, (b) selectively **targeted** to the site of transport, (c) vectorally **translocated** across the membrane, and (d) maintain a **permeability barrier** during these events. At the most fundamental level, these obstacles must have been solved in even the earliest life forms. This realization, together with the evolutionary relationship between the eukaryotic ER and bacterial plasma membrane, suggests a substantial conservation of the core principles of secretory protein translocation. Thus, assorted data using various model substrates from multiple systems (e.g., Bacteria, Archaea, yeast, and mammal)

*Ramanujan S. Hegde—Cell Biology and Metabolism Branch, NICHD, 18 Library Drive, Bldg. 18T, Room 101, National Institutes of Health, Bethesda, Maryland, U.S.A. Email: hegder@mail.nih.gov

Protein Movement Across Membranes, edited by Jerry Eichler. ©2005 Eurekah.com and Springer Science+Business Media.

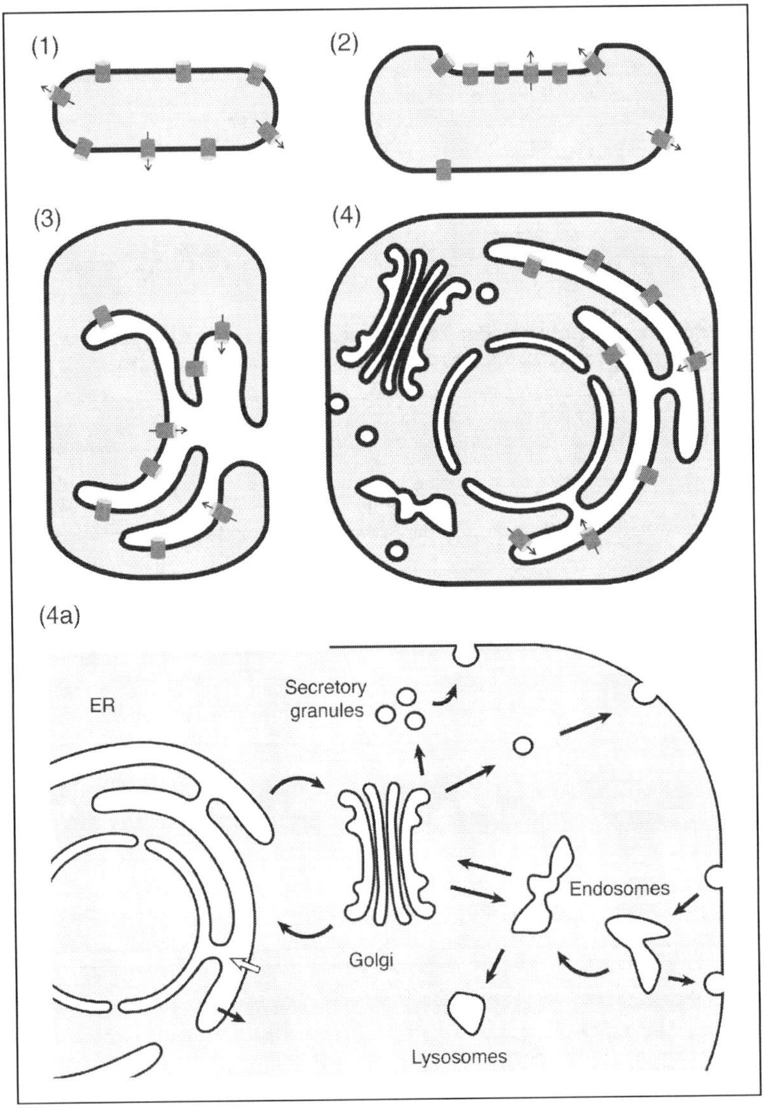

Figure 1. Evolution of the eukaryotic secretory pathway. Steps (1) through (4) depict successive stages in the generally accepted view of eukaryotic secretory pathway evolution from a prokaryotic ancestor. The cytoplasm is shown in gray, and translocons for protein secretion are depicted by cylinders with the direction of polypeptide transport indicated by an arrow. Note the relationship between secretion across the bacterial plasma membrane (in stage 1) and translocation into the ER (in stage 4). Diagram 4a shows a more detailed view of the mammalian secretory and endocytic pathways, with the primary pathways of protein traffic indicated by arrows. Essentially all of these pathways have been discovered to be regulated in a manner that allows some, but not other substrates to be trafficked in appropriate amounts to meet the changing demands of the cell. Notable examples include quality control at the ER, exit from the ER, sorting at the Golgi, regulated exocytosis, and endocytic sorting and degradation. By contrast, translocation into the ER (open arrow) is often regarded as a constitutive process where the presence of a signal sequence in a protein predetermines its entry into the ER.

and multiple approaches (biochemical, genetic, and structural) have often been consolidated into unifying models of protein translocation that are extrapolated to all systems.[1-4] While this provides a convenient framework for understanding protein translocation in general, it is apparent that further experiments will be required to either validate or revise the models for each individual system.

Basic Principles

Secretory and membrane proteins destined for the secretory pathway are recognized by the presence of hydrophobic domains in either signal sequences or transmembrane segments. N-terminal signal sequences (typically ~15-35 amino acids long) contain a hydrophobic core of at least 6 residues, while transmembrane segments have a hydrophobic stretch of between 16-25 residues. Aside from hydrophobicity, sequences used for the segregation of secretory and membrane proteins have no other features in common.[5,6] Indeed, the requirements are so degenerate that signals and transmembrane domains from prokaryotic and eukaryotic proteins are often functionally interchangeable,[7-9] and a surprising 20% of random sequences can at least partially mediate secretion from yeast.[10] Despite this tremendous diversity, signal sequences direct substrates into one of only two main translocation pathways in eukaryotes. In the cotranslational pathway (studied most extensively in the mammalian system), substrates are translocated across the membrane concurrent with their synthesis by membrane-bound ribosomes. In the post-translational pathway (studied primarily in the yeast system), the substrate is fully synthesized in the cytosol first, and translocated in a ribosome-independent fashion.

In cotranslational translocation, emergence from the ribosome of the first hydrophobic domain (either the signal sequence or transmembrane segment) allows its recognition in the cytosol by the signal recognition particle (SRP).[2,3] The complex of SRP and the ribosome-nascent chain (RNC) is then targeted to the membrane by an interaction with the SRP receptor (SR). At the membrane, the signal sequence is released by SRP, the RNC is transferred to the translocon, and the SRP-SR complex is dissociated. Thus, the targeting cycle culminates with delivery of the RNC to the translocon and recycling of components of the targeting machinery (SRP and SR) for the next substrate.

Nascent chains that are cotranslationally targeted to the translocon must then engage the translocation channel, mediate its opening, and be transported through it across the membrane. The central component of the translocation channel is the evolutionarily conserved heterotrimeric Sec61 complex.[11,12] The Sec61 complex, which has a high affinity for ribosomes,[13] provides a docking site for RNCs without the need for other components. However, docking of an RNC at the translocon is not sufficient to initiate translocation. Rather, engagement of the channel requires a functional signal sequence (or transmembrane domain), whose association with the Sec61 complex represents a second substrate recognition event during cotranslational translocation.[14]

This second recognition step may serve a 'proofreading' purpose to ensure that no non-signal-containing substrates that inadvertently target to the channel can engage it. More importantly, binding of the signal to the Sec61 complex triggers at least three essentially simultaneous changes in the RNC-translocon complex: (a) an increase in stability of the interaction between the RNC and translocon, (b) insertion of the nascent chain into the translocation channel, and (c) opening of the translocation channel towards the lumen.[14-18] Upon successful completion of these steps, the substrate resides in a continuous path running from the peptidyl transferase center in the ribosome, through the translocation channel, and into the ER lumen.[14,15,19] From this point, continued protein synthesis is thought to result in 'pushing' of the nascent chain through the channel and across the membrane.

Hence, the architecture of the RNC-translocon complex[20-22] biases the direction of nascent chain movement, thereby harnessing the energy of protein synthesis to simultaneously drive translocation.

Post-translational translocation operates in several qualitatively different ways. In eukaryotes, this pathway has been studied most extensively in yeast, where a seven protein Sec complex at the ER membrane and the lumenal chaperone BiP (known as Kar2p in yeast) have been identified as the essential translocation apparatus.[23-27] This Sec complex can be conceptually (and experimentally) divided into two sub-complexes: the trimeric Sec61 complex (homologous to the mammalian Sec61 complex), and the tetrameric Sec62/63 complex.[23] The Sec61 complex presumably forms a similar channel in the post-translational Sec complex as it does in the cotranslational translocon.[28] This means that the remaining components (the Sec62/63 subcomplex and BiP) must fulfill the functions otherwise provided in cotranslational translocation by the targeting machinery (SRP and SR) and ribosome, neither of which are involved in post-translational translocation.

Consistent with this idea, the Sec62/63 complex (but not BiP) is essential for signal sequence recognition by the Sec61 complex.[23,27,29] Thus, the Sec complex, by selectively binding signal-containing substrates, mediates targeting to the translocon in a single mechanistic step that replaces the series of targeting reactions involving the ribosome, SRP, SR, and translocon. Once substrate is bound to the Sec complex, the Sec61 translocation channel is thought to be engaged and opened in a similar fashion to the signal-mediated gating step in cotranslational translocation.[30] The substrate would then need to be moved unidirectionally through the Sec61 channel across the membrane.

Since vectorial movement of the substrate through the channel cannot exploit the energy of protein synthesis (as during cotranslational translocation), the actual transport step needs to occur differently. This function of biasing the direction of polypeptide movement is provided by BiP, a chaperone that binds the substrate on the lumenal side of the translocation channel to prevent its back-sliding into the cytosol.[23,25-27,31] Subsequent rounds of binding and release, stimulated by ATP hydrolysis, allows BiP to act as a molecular ratchet to drive substrate transport into the lumen.[32] The ATPase activity of BiP is regulated by Sec63p, a J-domain containing component of the Sec complex, which presumably also serves the function of recruiting BiP to the translocation channel.[27,33] Thus, the substrate is largely 'pulled' across the membrane from the lumenal side in the post-translational pathway, in contrast to being 'pushed' from the cytosolic side in cotranslational translocation.

A comparative analysis of these basic features of eukaryotic cotranslational and post-translational translocation reveals an important central theme (Fig. 2). It has become clear that the actual channel through which the polypeptide is translocated acts as a relatively passive conduit. It only acquires its functionality for substrate recognition and vectorial transport upon interaction with various binding partners. In cotranslational translocation, a key binding partner is the ribosome which acts to mediate translocon assembly, 'primes' the Sec61 complex for signal recognition, and couples the energy of protein synthesis to translocation. In post-translational translocation, the key binding partner is the Sec62/63 complex which, like the ribosome, facilitates translocon assembly, allows signal sequence recognition, and provides the driving force for translocation by recruiting and regulating the function of BiP at the translocation site. Indeed, even in the bacterial system, the homolog of the Sec61 complex (termed the SecY complex) interacts with the cytosolic SecA ATPase that both receives the substrate at the channel and drives its subsequent translocation across the membrane.[34] Thus, the highly conserved Sec61 channel can be exploited in several markedly different ways by various coassociating partners that mediate protein translocation across the eukaryotic ER or prokaryotic plasma membrane.[1,4]

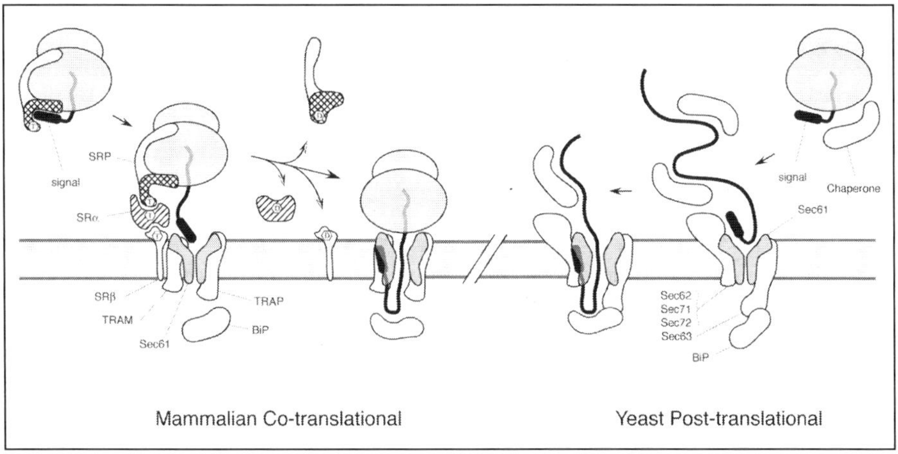

Figure 2. Pathways of ER protein translocation. The principal machinery and steps of the eukaryotic cotranslational and post-translational pathways are shown on the left and right, respectively. The components of each pathway that are conserved in all organisms (in both prokaryotes and eukaryotes) are shaded, and include the signal sequence, ribosome, SRP54 (along with a portion of its associated SRP RNA), SRα, and the Sec61 complex. Various other components that function in each pathway are also shown. The GTP- and GDP-bound states of the cotranslational targeting machinery are displayed with 'T' and 'D' respectively. The center two diagrams depict the comparable 'committed' stages of the two pathways to illustrate that in both, the Sec61 complex serves the same passive role as the channel while the associated components function to keep the polypeptide unfolded and move it vectorally into the lumen.

Molecular Details

Signal sequence recognition and targeting is understood in the greatest molecular detail for the cotranslational (i.e., SRP-dependent) pathway in eukaryotes. This is largely because the remarkable evolutionary conservation of this pathway from Bacteria to mammals has allowed the experimental results from multiple systems and approaches to be combined.[3] In higher eukaryotes, SRP is a ribonucleoprotein composed of six proteins (named by their apparent molecular weights: SRP72, SRP68, SRP54, SRP19, SRP14, and SRP9) and a ~300 nucleotide RNA (termed 7SL RNA or SRP RNA).[35,36] Of these components, SRP54 and a portion of the RNA are directly involved in both signal sequence recognition and the interaction with SR. Indeed, these two components define the minimal SRP that can be found in all organisms of every kingdom of life.[3] In almost all Bacteria, **only** these two components are found, indicating that they can perform all of the recognition and targeting functions necessary for translocation.[37-41]

Structural analysis of SRP54 homologues from several organisms[42-48] has revealed that it is universally organized into three functional segments: the M, N, and G domains. Of these, the M domain recognizes signal sequences via a deep, hydrophobic groove lined by the flexible side chains of several methionines. Phosphates of the RNA backbone are near one end of this groove, and may interact with basic residues that are often (but not always) adjacent to the hydrophobic core of signal sequences and transmembrane domains. These and other conserved features of SRP54 help to explain how it can accommodate a wide range of signal sequences whose only common feature is a hydrophobic segment, and why signals from diversely different organisms are often interchangeable.

In addition to signal sequence recognition, the other essential function of SRP is its interaction with SR to ensure the targeting of nascent secretory and membrane proteins to the translocon. The tight coordination of the series of interactions that imparts unidirectionality to the targeting phase of translocation is through the regulated GTPase activities of SRP and SR. The GTPase component of SRP resides in the G domain of SRP54 (see ref. 35). In eukaryotes, SR is a heterodimer of α and β subunits,[49] both of which are GTPases.[37,50] Of these, SRα is highly conserved from prokaryotes to mammals and, together with SRP54 and SRP RNA, represents the minimal targeting machinery found in all organisms.[37,41] Detailed mechanistic and structural analysis of this minimal SRP pathway, mostly using the model bacterial system, has revealed the essential aspects of their regulation during cotranslational targeting.

In the current working model, free SRP in the cytosol is in the GDP-bound state. Its association with the ribosome stimulates GTP binding,[51] and subsequent association with the signal sequence inhibits GTP hydrolysis.[52,53] Thus, the signal-SRP-ribosome ternary complex is likely to be in the GTP-bound state. Although less direct evidence exists for SRα, it is thought that its association with a vacant translocon at the membrane (directly in the case of prokaryotes, and indirectly via SRβ in eukaryotes) may similarly allow GTP binding and prevent GTP hydrolysis. Thus, the SR-translocon complex would also be in the GTP-bound state. The GTP-bound forms of SRP54 and SRα have a high affinity for each other,[53] allowing the delivery of signal-containing RNCs to the close proximity of an appropriately vacant translocon.[54,55]

The interaction between the GTPase domains of SRα and SRP54 stimulate the hydrolysis of GTP by each other (thereby acting as GTPase activating proteins, or GAPs, for one another).[56] The change in conformation that accompanies this GTP hydrolysis results in a weakening of the interaction between SRα and SRP54, allowing this complex to be dissociated for another round of targeting.[54,57,58] Many of the molecular details of this generally appealing scheme remain to be elucidated. For example, SRP RNA,[59,60] as well as the translocon[61,62] and the ribosome,[51] clearly facilitate aspects of SRP-SR interactions and their GTPase activities. However, the precise mechanisms remain elusive at the present time. The recently emerging wealth of structural information on SRP and SR should help to illuminate the molecular details of this framework.

Beyond these essential functions performed by the minimal components, the significantly more complex eukaryotic SRP and SR are likely to confer additional functionality and advantages to the cell. One such eukaryotic-specific feature is the slowing of translation upon signal sequence binding by SRP, a phenomenon termed 'elongation-arrest'.[63,64] The mechanism appears to involve occlusion of the elongation factor binding site on the ribosome by the SRP9 and SRP14 subunits of SRP.[65] The resulting decrease in translational rate serves to increase the time available for targeting to the translocation channel before excessive polypeptide synthesis precludes cotranslational transport. While translational attenuation by SRP is not essential for translocation,[66] it appears to be physiologically important under at least some growth conditions in vivo.[67] Whether the other subunits of SRP (SRP68, SRP72, and SRP19), each of which is important for assembly (particularly SRP19) and stability of the complete particle,[68] confer yet additional functionality to eukaryotic SRP remains largely unknown. Similarly, SRβ, a homolog for which does not exist in prokaryotes, is likely to provide the bridge that further regulates the coordinated transfer of RNCs from SRP to the translocon. This appears to be accomplished by the regulation of SRβ GTPase activity by both the ribosome[69] and the translocon,[62] with accompanying conformational changes that are suggested to affect the RNC-SRP54-SRα-SRβ interactions.[70]

Signal sequences and transmembrane domains are also recognized by the translocon at the membrane in all modes of translocation.[14,30,71-73] The purpose of this recognition is two-fold. First, it provides a mechanism for discriminating translocation substrates from other proteins. This is the sole discriminatory step in post-translational translocation, and a secondary (or 'proofreading') step in cotranslational translocation. Second, signal recognition by the translocon is essential for its opening (or gating) in preparation for substrate transport.[14,73] Since the core of the translocation channel in both co- and post-translational translocons of both prokaryotic and eukaryotic cells contains the Sec61 complex, the basic mechanism of signal recognition at the membrane in all cases is presumed (but not yet demonstrated directly) to be mechanistically similar. Hence, this step would appear to be a point of convergence for both co- and post-translational translocation pathways in different organisms. Indeed, recognition of signals (and transmembrane domains) by the Sec61 translocation channel is likely to be as ancient and evolutionarily conserved as signal recognition by SRP. Yet, the Sec61-mediated recognition step, by striking contrast to SRP-mediated recognition, is very poorly understood. This is in part because the membrane proteins involved in translocon signal recognition are significantly more difficult to manipulate and study, relative to the cytosolic SRP. However, if one is allowed some degree of extrapolation across species, a general framework and a few mechanistic details of signal recognition by the translocon can be compiled.

Cross-linking studies in both mammalian and yeast systems suggest that the signal sequence binds to a site that is at the interface of the Sec61 channel and the surrounding lipid bilayer.[30,71,74] Detailed analysis of the regions of yeast Sec61p that interact with the signal sequence of a model substrate (prepro-α-factor) has implicated transmembrane helices 2 and 7 as forming the binding site.[30] These same two helicies of the bacterial SecY complex were also observed to interact with a synthetic signal peptide in detergent solution.[72] All of these findings from the mammalian, yeast, and bacterial systems can now be reconciled with the crystal structure of an archaeal SecY complex.[75,76] This structure revealed that helicies 2 and 7 are indeed adjacent to each other and provide a lateral exit site from the proposed pore within SecY to the lipid bilayer.[76] Thus, it seems reasonable to conclude that in all systems, signal sequences (and transmembrane domains) of translocation substrates are recognized by a site in Sec61/SecY that is composed of the two transmembrane helicies (2 and 7) that line the lateral exit site from the translocation channel.

In addition to this 'generic' signal recognition site in Sec61, it is clear that other Sec61-associated components are also involved in signal recognition in many cases. These additional components may, directly or indirectly, stabilize signal sequence-Sec61 interactions for at least a subset of substrates. In cotranslational translocation, these additional components include the TRAM protein[77,78] and the tetrameric TRAP complex.[79] In both cases, these accessory translocon components are required in a signal sequence-dependent manner for the translocation of some, but not other substrates. TRAM has been directly implicated in interacting with the hydrophilic region that directly precedes the hydrophobic core of a signal peptide.[80] The role of TRAP is less clear, but it may act indirectly by stabilizing the Sec61 channel with which it directly interacts. In post-translational translocation, the Sec62/63 complex is absolutely required for signal sequence recognition by the Sec61 complex.[23] The mechanism is not yet clear, but it may be a combination of direct signal sequence interactions (e.g., with Sec62, which has been implicated in cross-linking studies[27,30,81]), or indirect effects as a consequence of stabilizing the Sec61 translocon.[28]

The features of the signal that determine the need for these additional components are not well-studied, nor are the mechanisms by which they facilitate recognition. Furthermore, whether yet other components are also involved in substrate-specific aspects of signal recognition is also not known. Numerous proteins, particularly in the mammalian system, have been identified to be at or near the site of translocation. These include proteins with known functions (such as the multi-protein oligosaccaryl transferase complex[82] or five protein signal peptidase complex[83]), as well as many others whose functions are not known.[84-87] While none of these are absolutely essential for translocation of at least the simplest model substrates,[12] it is not known whether they play essential or stimulatory roles in translocation of select substrates. As was exemplified by the TRAP complex, the functions of such accessory factors may elude detection[88] unless the proper substrate is examined.[79]

Maintaining the Membrane Permeability Barrier

During protein translocation, the membrane permeability barrier to the passage of small molecules should not be compromised. How this is achieved remains a matter of considerable debate. It is clear, however, that resolving this issue will require information about the architecture of the translocon, the structure of its individual constituents, and how they are assembled and changed during the functional translocation cycle. This will provide critical information about the nature of the translocation pore, its size, how it might be opened and closed, and how its permeability to small molecules can be controlled both during and in the absence of substrate translocation. At present, such structural and organizational information about the translocon and the pore are only beginning to emerge, leaving the mechanism of membrane permeability maintenance unresolved.

The first experimental studies to begin addressing the issues of pore size and membrane permeability were in the mammalian cotranslational system. In these experiments, translocation intermediates were assembled in which the substrate contained within it a fluorescently labeled amino acid at a defined position. The fluorophore was then used as a probe of both the environment surrounding the nascent chain[89] and the accessibility of this environment to exogenously added molecules capable of quenching the fluorophore.[15,90-92] The ability to control substrate length (and hence, the stage of translocation), the position of the probe, and the size and location of the fluorescence quenchers allowed various parameters of the translocon to be deduced. From these studies,[93] the pore sizes of inactive versus engaged translocons were measured to be ~8-10 Å and ~40-60 Å, respectively.[90] Preventing the passage of small molecules through this pore depended on alternately sealing the channel with either a ribosome on the cytosolic side or BiP on the lumenal side.[90-92] Sequences in the nascent polypeptide are proposed to choreograph the dynamics of channel gating by the ribosome and BiP to allow substrate transport without small molecule leakage.[92] Recently, an electrophysiological approach also suggested that purified Sec61 complex in lipid bilayers may contain pores as large as 60 Å that can be blocked by BiP.[94]

Although the model derived from the fluorescent probe approach is internally consistent and compatible with many other biochemical experiments in the mammalian cotranslational system, several arguments against it have been raised. In one experiment, the inability to detect folding of even a small domain while it is inside the translocon[95] seemed at odds with the proposed 40-60 Å pore size.[90,94] However, it is not clear how generalizable the results from either approach are since in each case, a single (and different) substrate has been examined to measure pore size. In other experiments, structural studies using cryo-electron microscopy (EM) of RNCs bound to the translocon failed to see a tight seal between the ribosome and translocation channel that was expected from the fluorescence quenching studies.[20-22,96] However, an inability to see density by cryo-EM can be difficult to interpret since it could be due to increased flexibility in those regions of the structure, loss of ancillary translocon components upon solubilization and sample preparation, or sample heterogeneity. Thus, cytosolic or membrane components in addition to the ones visualized by cryo-EM may form the putative seal between the ribosome and membrane. Indeed, several abundant membrane components have been identified associated with the translocon (some with large cytosolic domains such as p180)[84] whose functions remain unclear. Thus, there are some potentially plausible ways to reconcile much of the seemingly conflicting data gathered on membrane permeability and translocon architecture of the mammalian cotranslational system.

More problematic, however, is the argument that the proposed mechanism involving the ribosome and BiP during mammalian cotranslational translocation does not shed light on how the permeability problem is solved in other modes of translocation or in bacterial systems. In the post-translational pathway, the ribosome is not involved in translocation, precluding a role for it in maintaining the permeability barrier. In Bacteria, it is unclear what would serve the function of the lumenal gate proposed for BiP in the mammalian system. Because of these difficulties, a more generally applicable and evolutionarily conserved solution to the permeability barrier problem has been sought. The most insight into such a putatively conserved mechanism comes from interpretation of the recent high resolution crystal structure of an archaeal SecY complex.[75,76]

In this structure, a single SecY complex was found to form a channel-like structure with a very small pore flanked on the lumenal and cytosolic sides by funnels. The narrow constriction between these two funnels is only ~5-8 Å in diameter and lined by several hydrophobic residues that together form the 'pore ring.' If the channel formed by a single SecY complex is the functional pore through which the substrate is transported, the small size and flexibility of the 'pore ring' side chains would then form a relatively snug fit around a translocating polypeptide. This mechanism of translocation would solve the permeability problem because the nascent chain itself can occlude the channel during translocation. Furthermore, another small segment of the SecY protein (termed the 'plug' domain) appears to occlude the pore in its inactive state.[76] Thus, no additional components would be required to maintain permeability except the Sec61/SecY complex, which forms the channel in all modes of translocation.

The theoretical and experimental evidence that the translocation pore is indeed formed within a single SecY or Sec61 complex, despite its oligomerization into a larger structure,[22,28,97,98] is reviewed in detail elsewhere.[75,76] In essence, it is argued that a hydrophilic pore cannot be formed at the interface of multiple SecY complexes unless they 'face' each other, a configuration the authors of the structural work consider unlikely based on experiments examining the bacterial SecY complex.[99-102] Whether this proves to be true in all translocation systems, and hence explains the permeability problem, remains to be investigated. The alternative explanation is that in eukaryotic systems, the basic unit of translocation has evolved into a more malleable oligomeric structure in which the pores of multiple Sec61 complexes can indeed be combined to form a larger translocon that changes to meet the demands of the substrate.

This explanation would necessitate additional protein complexes that facilitate this reorganization and new mechanisms to solve the permeability problem. While this might seem unnecessarily complicated, it is not unreasonable given the existence of numerous eukaryotic-specific translocation components whose functions remain largely unknown (such as Sec62, Sec63, TRAM, or TRAP, among many others) At present, the choice among the different views depends largely on where a philosophical line is drawn. On the one hand is the tremendous degree of evolutionary conservation of the most fundamental features of protein translocation that has allowed information across multiple kingdoms to be combined into explanations applicable to all systems. On the other hand is the equally powerful feature of evolution to forge new biological principles using the same basic constituents. Clearly, the former is justified when one considers examples such as the SRP pathway, while the latter is strikingly exemplified by the evolution in eukaryotes of mechanisms to 'pull' nascent chains across the membrane from a system initially designed to 'push' such chains from the cytosolic side. Ultimately, experimental results will be needed to resolve these issues and determine the degree to which evolution has been conservative versus inventive in shaping eukaryotic protein translocation across the ER.

Regulation of Translocation

The evolution of a complex endomembrane system in eukaryotes (Fig. 1) provides several advantages to the cell, some of which are more obvious than others. These advantages include increased capacity, quality control, quantity control, and regulation. In this last section, examples of the ways in which the development of a multi-compartment secretory pathway has been exploited in complex eukaryotic organisms is discussed as a means of illustrating a general principle of regulated biological processes. This principle is then used to develop a rational framework for how and why eukaryotic protein translocation is likely to be a highly regulated process. And finally, the (albeit limited) data on translocational regulation is compiled into some potential mechansims by which ER protein translocation can be controlled in a substrate-specific manner.

The first advantage of compartmentalization, increased capacity, is a direct consequence of the substantially increased surface area of membrane across which a protein can be translocated. In the most extreme instance, the ER is expanded to almost completely fill the cells of highly secretory tissues such as the exocrine pancreas. The increased surface area (and hence capacity) conferred by the secretory and endocytic pathways is one reason (among many) that eukaryotic cells can be substantially larger than prokaryotes. The other three advantages are inter-related, and all a direct consequence of the fact that secretion in eukaryotes is a multi-step process that begins, not ends with translocation across the membrane. Thus, upon translocation, the protein is still available to a eukaryotic cell before its secretion, while in prokaryotes, translocation is largely synonymous with exit from the cell. This availability has been thoroughly exploited to confer several important advantages to eukaryotes.

The most important advantage is the opportunity for quality and quantity control: since a translocated protein in eukaryotes is not lost to the extracellular space, there is time to impose a 'recall' in instances where the protein is not desired. Hence, if a protein is not matured or assembled properly, it is rerouted for degradation (i.e., quality control),[103,104] thereby avoiding the potentially detrimental consequences of misfolded or incomplete secretory and membrane proteins. This has almost certainly facilitated the evolution of very complex secretory proteins (such as apolipoprotein B) or multi-component membrane protein complexes (such as the T-cell receptor). Similarly, regulated degradation after translocation allows the abundance of secretory or membrane proteins to be modulated in response to need (i.e., quantity control, exemplified by HMG-CoA reductase[105] or apolipoprotein B[106]). Furthermore, the intracellular compartmentalization of secretion allows secretory and membrane proteins to be stored until they are needed,[107] at which point they can be rapidly delivered to selected regions of the cell surface by exocytosis. Thus, secretion of extracellular proteins or surface expression of membrane proteins can be rapid, quantal, and temporally and spatially regulated. These examples illustrate an important general principle: the disadvantages of increased cost and lower efficiency of a more complex, multi-step process (e.g., the secretory pathway) can be offset by the benefits of a greater degree of regulatory control. Thus, potentially regulatory aspects of the secretory pathway are likely to be most thoroughly developed in systems where control, and not just energetic cost, is of the utmost importance.

In which organisms is the highest premium placed on precise control of secretory and membrane protein biogenesis? The answer is multicellular organisms, whose fitness depends not only on the health of individual cells, but equally (or perhaps even more) on the ways those cells interact, communicate, and function as complex units. Such communication and interactions are intimately dependent on secreted and cell surface proteins whose amounts at the right time and place must be carefully regulated. Thus, completely healthy **individual** cells in a complex organism can nonetheless lead to failure of the organism if they do not function coordinately in extremely precise ways. Countless examples of this idea can be found in human physiology and disease, including the regulation of blood pressure, reproductive cycles, stress, appetite, and weight regulation. It should, therefore, come as no surprise that **each and every step** in the secretory pathway that has been examined was discovered to be regulated to tightly control the levels of secretory and membrane proteins in response to cellular and organismal needs (Fig. 1). Will protein translocation prove to be any different, once more complex (and subtle) aspects of this process have received experimental attention? Almost certainly not.

How then might one conceptualize a framework for translocational regulation that can guide future investigation? At the outset, it is instructive to consider analogies to other regulatory systems for common themes that can be applied to translocation. In this vein, a grossly simplified discussion of transcriptional promoters and their regulation is useful[108-111] (although similar arguments can be made equally well with any other regulatory process). In transcription, sequence features that are common to all promoters are accompanied by sequence elements that are unique to each individual promoter.[112] Thus, each promoter is unique, but contains at least some common elements that allow it to be recognized as a promoter per se. The common elements allow a core (or 'general') machinery to mediate transcription,[108,109] while the unique elements impose requirements for additional machinery that regulate the recruitment or activity of the core components.[110,111] The combinatorial expression or modification of the unique machinery can dramatically influence the activity of any given promoter. By regulating individual components of the unique machinery in a temporal or cell type-specific manner, transcriptional regulation of individual promoters can be achieved independently of each other. Thus, sequence diversity of promoters combined with diversity in the components that recognize them allows selective regulation of genes that all, nonetheless, use a commonly shared core machinery for transcription.

Applying this general idea to translocation allows at least one mechanism of regulation to be conceptualized. Here, signal sequences are viewed as loosely analogous to promoters, and the evolutionarily conserved components of the translocation machinery (i.e., SRP, SR, and Sec61 complexes) are analogous to the core transcriptional machinery. Signal sequences are indeed extremely diverse, with each substrate containing an effectively unique signal, while nonetheless sharing certain common, recognizable features.[5,6] The common features of the signal appear to be the elements that are recognized by the core machinery, such as SRP54 and the Sec61 complex. The unique features of the signal appear to impose additional constraints on signal function by requiring the presence of additional factors at the translocation site such as TRAM or the TRAP complex.[77-79] These additional components can be modified (e.g., by phosphorylation[113-115]), which potentially may selectively modulate their activity (although this has yet to be examined). Thus, even using only the limited information that is currently known, one can easily envision the basic elements of a substrate-specific system of translocational regulation (Fig. 3): (a) diversity in structure and function of signal sequences that share a bare minimum of common features, (b) diversity in 'accessory' components that influence recognition by a core translocation machinery of some, but not other signals, and (c) selective changes in expression or modification of the 'accessory' components that could affect the outcome of translocation for some, but not other substrates.

This view of regulating translocation by the combinatorial functions of accessory components can be readily expanded to incorporate the many other factors at or near the site of translocation whose functions remain elusive. In the mammalian system, these include Sec62, Sec63, p180, p34, a TRAM homolog, and yet unidentified proteins observed by cross-linking studies. Each of these components could potentially play stimulatory (or inhibitory) roles in the translocation of selected substrates, with the specificity encoded in the sequence diversity of the signal. Such accessory components can not only be modified, but themselves regulated at steps such as alternative splicing[116] or differential expression[117] to influence their function. Thus, there exist more than enough sources for modulatory activities to theoretically provide exquisite specificity in the regulation of signal sequence function, and hence translocation.

Initial evidence that protein translocation can indeed be modulated in a substrate-selective, cell-type specific way has recently been provided by quantitatively examining the efficiency of signal sequence function in vivo.[118] Not only were different signal sequences found to have different efficiencies within a given cell type, but they also varied independently in a cell type-specific manner. For example, one signal sequence was observed to be significantly more efficient than another signal in a particular type of cell; in a different cell type, the two signals were found to be equally inefficient. Thus, the entry of proteins into the ER is not necessarily

A

Protein	Signal sequence	Length	Charge
BiP	MKLSLVAAMLLLLSAARA	18	+1
Prolactin	MNIKGSPWKGSLLLLLVSNLLLCQSVAP	28	+2
TGF-β2	MHYCVLSAFLILHLVTVAL	19	0
TGF-β3	MKMHLQRALVVLALLNFATVSLS	23	+2
Growth hormone	MATGSRTSLLLAFGLLCLPWLQEGSA	26	+1
Interferonγ	MKYTSYILAFQLCIVLG	17	+1
Osteopontin	MRIAVICFCLLGITCA	16	+1
Leptin	MHWGTLCGFLWLWPYLFYVQA	21	0
ANP	MSSFSTTTVSFLLLLAFQL	19	0
Apo-A1	MKAAVLTLAVLFLTGSQA	18	+1
EGF-receptor	MRPSGTAGAALLALLAALCPRA	24	+1
Glucagon	MKSIYFVAGLFVMLVQG	19	+1
Inhibinβ	MPLLWLRGFLLASCWIIVRSSPTPGS	27	0
HLA-DCβ	MSWKKALRIPGGLRVATVTLMLAMLSTPVAEG	32	+2
Choriogonadotropin	MEMFQGLLLLLLLSMGGTWA	20	-1

B

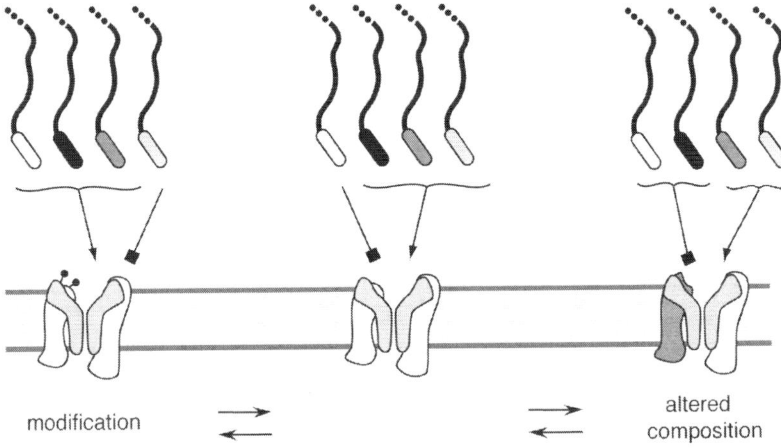

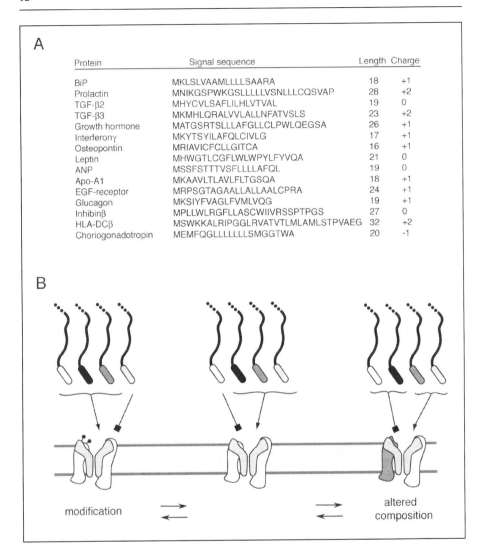

Figure 3. Conceptual framework for translocational regulation. One potential mechanism for regulating the translocation of secretory and membrane proteins into the eukaryotic ER is shown. Here, the sequence diversity of signals can be exploited to selectively influence the translocation efficiencies, and hence functional expression, of some but not other secretory pathway proteins. The key discriminatory step is proposed to occur at the translocon, whose modification or composition would affect its selectivity for the range of signal sequences that it can recognize to initiate translocation. Thus, the encoding of a signal sequence in a protein is not viewed as a guarantee of its entry into the secretory pathway; rather, the signal is a provisional 'license' for translocation that is contingent on the translocon, whose functional state can be modified in response to cellular need. Panel A shows a representative sample of several signal sequences from human secretory proteins. Note the wide diversity of amino acid sequence, composition, length, and charge of the N-terminal domain preceding the hydrophobic sequence. Panel B illustrates the concept of translocational regulation described above. In this example, several substrates that differ in their signal sequences are either accepted (arrow) or rejected (square) by each of the hypothetical translocons. Note that the selectivity is altered upon changes in either the modification state (left) or composition (right) of the translocon machinery.

a constitutive process predestined by the sequence of the substrate. Rather, it is dependent on and potentially regulated by the machinery that mediates its translocation.

Consistent with this view are the numerous examples in which secretory pathway proteins are also found in alternative locations (summarized in ref. 118), the extent of which can vary in a cell type-specific manner. The mechanistic basis or physiologic relevance of these observations is not yet clear, but at least some of these examples are likely to be a consequence of translocational regulation. Clearly, modulating translocation not only provides one means of quantity control (i.e., the ability to change the abundance of the protein in the secretory pathway), but also a mechanism to generate an alternative form of the same protein in another compartment, where it could potentially serve a second function. Examples of proteins that may have such alternative functions in different compartments have been suggested (see ref. 118 for a summary). The degree to which translocational regulation is beneficially utilized for the generation of functional diversity or quantity control of secretory pathway proteins remains to be investigated.

Conclusions

The eukaryotic translocation pathways contain a very well-recognizable core machinery whose identity and mechanisms of action are remarkably well-conserved across all kingdoms of life. The signal sequences and transmembrane domains that engage this machinery are essentially indistinguishable when comparing the prokaryotic versus eukaryotic populations of substrates. Despite these similarities in substrate clientele, the core translocation machinery has been embellished at every conceivable step during the evolution of eukaryotes (Fig. 2). The reasons for this increase in complexity, which are probably varied and numerous, remain poorly understood. One general explanation may be that a higher premium is placed in more complex organisms on fidelity of protein biogenesis. Not only are proteins generally more complex and multi-domained in eukaryotes, but the consequences of their misfolding may be more detrimental in cells whose proper function depends on a larger set of intersecting biochemical pathways. This may be particularly important in highly differentiated cells of multi-cellular organisms that must live long periods of time without replacement by cell division. Thus, very tight control of targeting and translocation, with contingencies for errors at each step, may provide subtle advantages to eukaryotes that outweigh the costs of increased complexity and energy expenditure. A more important reason for the multiple layers of complexity during eukaryotic translocation may be to facilitate cellular control at each step. This general theme of embellishing a basic process to allow for regulation is seen in virtually every other cellular process such as transcription, translation, or cell division. By adding accessory components whose activities can be used to modulate a core machinery, a biological process such as transcription can be changed in response to cellular demand or environmental conditions. If and how protein translocation can be regulated remains essentially unexplored at the present time, but is envisioned to utilize themes common to other biological regulatory processes. This concept of translocational regulation represents a fertile area for future research in this field.

References

1. Pohlschroder M, Prinz WA, Hartmann E et al. Protein translocation in the three domains of life: Variations on a theme. Cell 1997; 91:563-566.
2. Walter P, Johnson AE. Signal sequence recognition and protein targeting to the endoplasmic reticulum membrane. Annu Rev Cell Biol 1994; 10:87-119.
3. Keenan RJ, Freymann DM, Stroud RM et al. The signal recognition particle. Annu Rev Biochem 2001; 70:755-775.
4. Rapoport TA, Jungnickel B, Kutay U. Protein transport across the eukaryotic endoplasmic reticulum and bacterial inner membranes. Annu Rev Biochem 1996; 65:271-303.

5. Randall LL, Hardy SJ. Unity in function in the absence of consensus in sequence: Role of leader peptides in export. Science 1989; 243:1156-1159.
6. von Heijne G. Signal sequences. The limits of variation. J Mol Biol 1985; 184:99-105.
7. Lingappa VR, Chaidez J, Yost CS et al. Determinants for protein localization: Beta-lactamase signal sequence directs globin across microsomal membranes. Proc Natl Acad Sci USA 1984; 81:456-460.
8. Wiedmann M, Huth A, Rapoport TA. Xenopus oocytes can secrete bacterial beta-lactamase. Nature 1984; 309:637-639.
9. Hall J, Hazlewood GP, Surani MA et al. Eukaryotic and prokaryotic signal peptides direct secretion of a bacterial endoglucanase by mammalian cells. J Biol Chem 1990; 265:19996-19999.
10. Kaiser CA, Preuss D, Grisafi P et al. Many random sequences functionally replace the secretion signal sequence of yeast invertase. Science 1987; 235:312-317.
11. Gorlich D, Prehn S, Hartmann E et al. A mammalian homolog of SEC61p and SECYp is associated with ribosomes and nascent polypeptides during translocation. Cell 1992; 71:489-503.
12. Gorlich D, Rapoport TA. Protein translocation into proteoliposomes reconstituted from purified components of the endoplasmic reticulum membrane. Cell 1993; 75:615-630.
13. Kalies KU, Gorlich D, Rapoport TA. Binding of ribosomes to the rough endoplasmic reticulum mediated by the Sec61p-complex. J Cell Biol 1994; 126:925-934.
14. Jungnickel B, Rapoport TA. A posttargeting signal sequence recognition event in the endoplasmic reticulum membrane. Cell 1995; 82:261-270.
15. Crowley KS, Liao S, Worrell VE et al. Secretory proteins move through the endoplasmic reticulum membrane via an aqueous, gated pore. Cell 1994; 78:461-471.
16. Wolin SL, Walter P. Discrete nascent chain lengths are required for the insertion of presecretory proteins into microsomal membranes. J Cell Biol 1993; 121:1211-1219.
17. Matlack KE, Walter P. The 70 carboxyl-terminal amino acids of nascent secretory proteins are protected from proteolysis by the ribosome and the protein translocation apparatus of the endoplasmic reticulum membrane. J Biol Chem 1995; 270:6170-6180.
18. Murphy 3rd EC, Zheng T, Nicchitta CV. Identification of a novel stage of ribosome/nascent chain association with the endoplasmic reticulum membrane. J Cell Biol 1997; 136:1213-1226.
19. Kim SJ, Mitra D, Salerno JR et al. Signal sequences control gating of the protein translocation channel in a substrate-specific manner. Dev Cell 2002; 2:207-217.
20. Beckmann R, Bubeck D, Grassucci R et al. Alignment of conduits for the nascent polypeptide chain in the ribosome-Sec61 complex. Science 1997; 278:2123-2126.
21. Menetret JF, Neuhof A, Morgan DG et al. The structure of ribosome-channel complexes engaged in protein translocation. Mol Cell 2000; 6:1219-1232.
22. Beckmann R, Spahn CM, Eswar N et al. Architecture of the protein-conducting channel associated with the translating 80S ribosome. Cell 2001; 107:361-372.
23. Panzner S, Dreier L, Hartmann E et al. Posttranslational protein transport in yeast reconstituted with a purified complex of Sec proteins and Kar2p. Cell 1995; 81:561-570.
24. Deshaies RJ, Sanders SL, Feldheim DA et al. Assembly of yeast Sec proteins involved in translocation into the endoplasmic reticulum into a membrane-bound multisubunit complex. Nature 1991; 349:806-808.
25. Sanders SL, Whitfield KM, Vogel JP et al. Sec61p and BiP directly facilitate polypeptide translocation into the ER. Cell 1992; 69:353-365.
26. Brodsky JL, Schekman R. A Sec63p-BiP complex from yeast is required for protein translocation in a reconstituted proteoliposome. J Cell Biol 1993; 123:1355-1363.
27. Lyman SK, Schekman R. Binding of secretory precursor polypeptides to a translocon subcomplex is regulated by BiP. Cell 1997; 88:85-96.
28. Hanein D, Matlack KE, Jungnickel B et al. Oligomeric rings of the Sec61p complex induced by ligands required for protein translocation. Cell 1996; 87:721-732.
29. Matlack KE, Plath K, Misselwitz B et al. Protein transport by purified yeast Sec complex and Kar2p without membranes. Science 1997; 277:938-941.
30. Plath K, Mothes W, Wilkinson BM et al. Signal sequence recognition in posttranslational protein transport across the yeast ER membrane. Cell 1998; 94:795-807.

31. Lyman SK, Schekman R. Interaction between BiP and Sec63p is required for the completion of protein translocation into the ER of Saccharomyces cerevisiae. J Cell Biol 1995; 131:1163-1171.
32. Matlack KE, Misselwitz B, Plath K et al. BiP acts as a molecular ratchet during posttranslational transport of prepro-alpha factor across the ER membrane. Cell 1999; 97:553-564.
33. Misselwitz B, Staeck O, Matlack KE et al. Interaction of BiP with the J-domain of the Sec63p component of the endoplasmic reticulum protein translocation complex. J Biol Chem 1999; 274:20110-20115.
34. Duong F, Eichler J, Price A et al. Biogenesis of the gram-negative bacterial envelope. Cell 1997; 91:567-573.
35. Walter P, Blobel G. Disassembly and reconstitution of signal recognition particle. Cell 1983; 34:525-533.
36. Hann BC, Walter P. The signal recognition particle in S. cerevisiae. Cell 1991; 67:131-144.
37. Bernstein HD, Poritz MA, Strub K et al. Model for signal sequence recognition from amino-acid sequence of 54K subunit of signal recognition particle. Nature 1989; 340:482-486.
38. Poritz MA, Bernstein HD, Strub K et al. An E. coli ribonucleoprotein containing 4.5S RNA resembles mammalian signal recognition particle. Science 1990; 250:1111-1117.
39. Powers T, Walter P. Cotranslational protein targeting catalyzed by the Escherichia coli signal recognition particle and its receptor. EMBO J 1997; 16:4880-4886.
40. Bernstein HD, Zopf D, Freymann DM et al. Functional substitution of the signal recognition particle 54-kDa subunit by its Escherichia coli homolog. Proc Natl Acad Sci USA 1993; 90:5229-5233.
41. Dobberstein B. Protein transport. On the beaten pathway. Nature 1994; 367:599-600.
42. Keenan RJ, Freymann DM, Walter P et al. Crystal structure of the signal sequence binding subunit of the signal recognition particle. Cell 1998; 94:181-191.
43. Freymann DM, Keenan RJ, Stroud RM et al. Structure of the conserved GTPase domain of the signal recognition particle. Nature 1997; 385:361-364.
44. Rosendal KR, Wild K, Montoya G et al. Crystal structure of the complete core of archaeal signal recognition particle and implications for interdomain communication. Proc Natl Acad Sci USA 2003; 100:14701-14706.
45. Montoya G, Svensson C, Luirink J et al. Crystal structure of the NG domain from the signal-recognition particle receptor FtsY. Nature 1997; 385:365-368.
46. Clemons Jr WM, Gowda K, Black SD et al. Crystal structure of the conserved subdomain of human protein SRP54M at 2.1 A resolution: Evidence for the mechanism of signal peptide binding. J Mol Biol 1999; 292:697-705.
47. Doudna JA, Batey RT. Structural insights into the signal recognition particle. Annu Rev Biochem 2004; 73:539-557.
48. Batey RT, Rambo RP, Lucast L et al. Crystal structure of the ribonucleoprotein core of the signal recognition particle. Science 2000; 287:1232-1239.
49. Tajima S, Lauffer L, Rath VL et al. The signal recognition particle receptor is a complex that contains two distinct polypeptide chains. J Cell Biol 1986; 103:1167-1178.
50. Miller JD, Tajima S, Lauffer L et al. The beta subunit of the signal recognition particle receptor is a transmembrane GTPase that anchors the alpha subunit, a peripheral membrane GTPase, to the endoplasmic reticulum membrane. J Cell Biol 1995; 128:273-282.
51. Bacher G, Lutcke H, Jungnickel B et al. Regulation by the ribosome of the GTPase of the signal-recognition particle during protein targeting. Nature 1996; 381:248-251.
52. Miller JD, Wilhelm H, Gierasch L et al. GTP binding and hydrolysis by the signal recognition particle during initiation of protein translocation. Nature 1993; 366:351-354.
53. Miller JD, Bernstein HD, Walter P. Interaction of E. coli Ffh/4.5S ribonucleoprotein and FtsY mimics that of mammalian signal recognition particle and its receptor. Nature 1994; 367:657-659.
54. Connolly T, Rapiejko PJ, Gilmore R. Requirement of GTP hydrolysis for dissociation of the signal recognition particle from its receptor. Science 1991; 252:1171-1173.
55. Connolly T, Gilmore R. The signal recognition particle receptor mediates the GTP-dependent displacement of SRP from the signal sequence of the nascent polypeptide. Cell 1989; 57:599-610.
56. Powers T, Walter P. Reciprocal stimulation of GTP hydrolysis by two directly interacting GTPases. Science 1995; 269:1422-1424.

57. Egea PF, Shan SO, Napetschnig J et al. Substrate twinning activates the signal recognition particle and its receptor. Nature 2004; 427:215-221.
58. Focia PJ, Shepotinovskaya IV, Seidler JA et al. Heterodimeric GTPase core of the SRP targeting complex. Science 2004; 303:373-377.
59. Peluso P, Shan SO, Nock S et al. Role of SRP RNA in the GTPase cycles of Ffh and FtsY. Biochemistry 2001; 40:15224-15233.
60. Peluso P, Herschlag D, Nock S et al. Role of 4.5S RNA in assembly of the bacterial signal recognition particle with its receptor. Science 2000; 288:1640-1643.
61. Song W, Raden D, Mandon EC et al. Role of Sec61alpha in the regulated transfer of the ribosome-nascent chain complex from the signal recognition particle to the translocation channel. Cell 2000; 100:333-343.
62. Fulga TA, Sinning I, Dobberstein B et al. SRbeta coordinates signal sequence release from SRP with ribosome binding to the translocon. EMBO J 2001; 20:2338-2347.
63. Walter P, Blobel G. Translocation of proteins across the endoplasmic reticulum III. Signal recognition protein (SRP) causes signal sequence-dependent and site-specific arrest of chain elongation that is released by microsomal membranes. J Cell Biol 1981; 91:557-561.
64. Wolin SL, Walter P. Signal recognition particle mediates a transient elongation arrest of preprolactin in reticulocyte lysate. J Cell Biol 1989; 109:2617-2622.
65. Halic M, Becker T, Pool MR et al. Structure of the signal recognition particle interacting with the elongation-arrested ribosome. Nature 2004; 427:808-814.
66. Siegel V, Walter P. Elongation arrest is not a prerequisite for secretory protein translocation across the microsomal membrane. J Cell Biol 1985; 100:1913-1921.
67. Mason N, Ciufo LF, Brown JD. Elongation arrest is a physiologically important function of signal recognition particle. EMBO J 2000; 19:4164-4174.
68. Brown JD, Hann BC, Medzihradszky KF et al. Subunits of the Saccharomyces cerevisiae signal recognition particle required for its functional expression. EMBO J 1994; 13:4390-4400.
69. Bacher G, Pool M, Dobberstein B. The ribosome regulates the GTPase of the beta-subunit of the signal recognition particle receptor. J Cell Biol 1999; 146:723-730.
70. Schwartz T, Blobel G. Structural basis for the function of the beta subunit of the eukaryotic signal recognition particle receptor. Cell 2003; 112:793-803.
71. Mothes W, Jungnickel B, Brunner J et al. Signal sequence recognition in cotranslational translocation by protein components of the endoplasmic reticulum membrane. J Cell Biol 1998; 142:355-364.
72. Wang L, Miller A, Rusch SL et al. Demonstration of a specific Escherichia coli SecY-signal peptide interaction. Biochemistry 2004; 43:13185-13192.
73. Simon SM, Blobel G. Signal peptides open protein-conducting channels in E. coli. Cell 1992; 69:677-684.
74. Martoglio B, Hofmann MW, Brunner J et al. The protein-conducting channel in the membrane of the endoplasmic reticulum is open laterally toward the lipid bilayer. Cell 1995; 81:207-214.
75. Clemons Jr WM, Menetret JF, Akey CW et al. Structural insight into the protein translocation channel. Curr Opin Struct Biol 2004; 14:390-396.
76. Van den Berg B, Clemons Jr WM, Collinson I et al. X-ray structure of a protein-conducting channel. Nature 2004; 427:36-44.
77. Gorlich D, Hartmann E, Prehn S et al. A protein of the endoplasmic reticulum involved early in polypeptide translocation. Nature 1992; 357:47-52.
78. Voigt S, Jungnickel B, Hartmann E et al. Signal sequence-dependent function of the TRAM protein during early phases of protein transport across the endoplasmic reticulum membrane. J Cell Biol 1996; 134:25-35.
79. Fons RD, Bogert BA, Hegde RS. Substrate-specific function of the translocon-associated protein complex during translocation across the ER membrane. J Cell Biol 2003; 160:529-539.
80. High S, Martoglio B, Gorlich D et al. Site-specific photocross-linking reveals that Sec61p and TRAM contact different regions of a membrane-inserted signal sequence. J Biol Chem 1993; 268:26745-26751.
81. Plath K, Wilkinson BM, Stirling CJ et al. Interactions between Sec complex and prepro-alpha-factor during posttranslational protein transport into the endoplasmic reticulum. Mol Biol Cell 2004; 15:1-10.

82. Dempski Jr RE, Imperiali B. Oligosaccharyl transferase: Gatekeeper to the secretory pathway. Curr Opin Chem Biol 2002; 6:844-850.
83. Evans EA, Gilmore R, Blobel G. Purification of microsomal signal peptidase as a complex. Proc Natl Acad Sci USA 1986; 83:581-585.
84. Savitz AJ, Meyer DI. 180-kD ribosome receptor is essential for both ribosome binding and protein translocation. J Cell Biol 1993; 120:853-863.
85. Meyer HA, Grau H, Kraft R et al. Mammalian Sec61 is associated with Sec62 and Sec63. J Biol Chem 2000; 275:14550-14557.
86. Tyedmers J, Lerner M, Bies C et al. Homologs of the yeast Sec complex subunits Sec62p and Sec63p are abundant proteins in dog pancreas microsomes. Proc Natl Acad Sci USA 2000; 97:7214-7219.
87. Tazawa S, Unuma M, Tondokoro N et al. Identification of a membrane protein responsible for ribosome binding in rough microsomal membranes. J Biochem (Tokyo) 1991; 109:89-98.
88. Migliaccio G, Nicchitta CV, Blobel G. The signal sequence receptor, unlike the signal recognition particle receptor, is not essential for protein translocation. J Cell Biol 1992; 117:15-25.
89. Crowley KS, Reinhart GD, Johnson AE. The signal sequence moves through a ribosomal tunnel into a noncytoplasmic aqueous environment at the ER membrane early in translocation. Cell 1993; 73:1101-1115.
90. Hamman BD, Chen JC, Johnson EE et al. The aqueous pore through the translocon has a diameter of 40-60 A during cotranslational protein translocation at the ER membrane. Cell 1997; 89:535-544.
91. Hamman BD, Hendershot LM, Johnson AE. BiP maintains the permeability barrier of the ER membrane by sealing the lumenal end of the translocon pore before and early in translocation. Cell 1998; 92:747-758.
92. Liao S, Lin J, Do H et al. Both lumenal and cytosolic gating of the aqueous ER translocon pore are regulated from inside the ribosome during membrane protein integration. Cell 1997; 90:31-41.
93. Johnson AE, van Waes MA. The translocon: A dynamic gateway at the ER membrane. Annu Rev Cell Dev Biol 1999; 15:799-842.
94. Wirth A, Jung M, Bies C et al. The Sec61p complex is a dynamic precursor activated channel. Mol Cell 2003; 12:261-268.
95. Kowarik M, Kung S, Martoglio B et al. Protein folding during cotranslational translocation in the endoplasmic reticulum. Mol Cell 2002; 10:769-778.
96. Morgan DG, Menetret JF, Neuhof A et al. Structure of the mammalian ribosome-channel complex at 17A resolution. J Mol Biol 2002; 324:871-886.
97. Snapp EL, Reinhart GA, Bogert BA et al. The organization of engaged and quiescent translocons in the endoplasmic reticulum of mammalian cells. J Cell Biol 2004; 164:997-1007.
98. Meyer TH, Menetret JF, Breitling R et al. The bacterial SecY/E translocation complex forms channel-like structures similar to those of the eukaryotic Sec61p complex. J Mol Biol 1999; 285:1789-1800.
99. Breyton C, Haase W, Rapoport TA et al. Three-dimensional structure of the bacterial protein-translocation complex SecYEG. Nature 2002; 418:662-665.
100. Duong F. Binding, activation and dissociation of the dimeric SecA ATPase at the dimeric SecYEG translocase. EMBO J 2003; 22:4375-4384.
101. Kaufmann A, Manting EH, Veenendaal AK et al. Cysteine-directed cross-linking demonstrates that helix 3 of SecE is close to helix 2 of SecY and helix 3 of a neighboring SecE. Biochemistry 1999; 38:9115-9125.
102. van der Sluis EO, Nouwen N, Driessen AJ. SecY-SecY and SecY-SecG contacts revealed by site-specific crosslinking. FEBS Lett 2002; 527:159-165.
103. Ellgaard L, Helenius A. Quality control in the endoplasmic reticulum. Nat Rev Mol Cell Biol 2003; 4:181-191.
104. Tsai B, Ye Y, Rapoport TA. Retro-translocation of proteins from the endoplasmic reticulum into the cytosol. Nat Rev Mol Cell Biol 2002; 3:246-255.
105. Hampton RY. Proteolysis and sterol regulation. Annu Rev Cell Dev Biol 2002; 18:345-378.
106. Davidson NO, Shelness GS. Apolipoprotein B: mRNA editing, lipoprotein assembly, and presecretory degradation. Annu Rev Nutr 2000; 20:169-193.

107. Arvan P, Castle D. Sorting and storage during secretory granule biogenesis: Looking backward and looking forward. Biochem J 1998; 332:593-610.
108. Roeder RG. The role of general initiation factors in transcription by RNA polymerase II. Trends Biochem Sci 1996; 21:327-335.
109. Novina CD, Roy AL. Core promoters and transcriptional control. Trends Genet 1996; 12:351-355.
110. Kornberg RD. RNA polymerase II transcription control. Trends Biochem Sci 1996; 21:325-326.
111. Verrijzer CP, Tjian R. TAFs mediate transcriptional activation and promoter selectivity. Trends Biochem Sci 1996; 21:338-342.
112. Fickett JW, Hatzigeorgiou AG. Eukaryotic promoter recognition. Genome Res 1997; 7:861-878.
113. Prehn S, Herz J, Hartmann E et al. Structure and biosynthesis of the signal-sequence receptor. Eur J Biochem 1990; 188:439-445.
114. Gruss OJ, Feick P, Frank R et al. Phosphorylation of components of the ER translocation site. Eur J Biochem 1999; 260:785-793.
115. Ou WJ, Thomas DY, Bell AW et al. Casein kinase II phosphorylation of signal sequence receptor alpha and the associated membrane chaperone calnexin. J Biol Chem 1992; 267:23789-23796.
116. Noel P, Cartwright IL. A Sec62p-related component of the secretory protein translocon from Drosophila displays developmentally complex behavior. EMBO J 1994; 13:5253-5261.
117. Holthuis JC, van Riel MC, Martens GJ. Translocon-associated protein TRAP delta and a novel TRAP-like protein are coordinately expressed with pro-opiomelanocortin in Xenopus intermediate pituitary. Biochem J 1995; 312:205-213.
118. Levine CG, Mitra D, Sharma A et al. The efficiency of protein compartmentalization into the secretory pathway. Mol Biol Cell 2005; 16:279-291.

Chapter 2

Preprotein Translocation through the Sec Translocon in Bacteria

Antoine P. Maillard, Kenneth K.Y. Chan and Franck Duong*

Abstract

The Sec translocase or translocon is the essential and ubiquitous system for protein translocation across or into the membrane. The core channel, the SecYE complex, is conserved across biological kingdoms and most of the polypeptide chains which are routed to extracellular or membrane locations in Bacteria use this pathway. Biochemical and genetic approaches have yielded a substantial body of information about functional aspects of Sec-mediated translocation and this information has recently been enriched with structural data at atomic resolution. This chapter reviews previously acquired facts and concepts concerning the Sec translocase of Bacteria in light of recent structural results and considers implications of these findings.

Introduction

In *E. coli*, components of the Sec pathway were identified during the mid-1980s using elegant genetic screens. Conditional-lethal mutations associated with a generalized protein-secretion defect or mutations restoring translocation of proteins with secretion-defective leader peptides allowed the identification of most of the Sec components.[1-3] The translocation pathway was then successfully reconstituted in vitro in the early 1990s to allow biochemical dissection of the subreactions of the translocation event.[4] It was shown that targeting of the protein substrate to the translocase is mediated by the dedicated chaperone SecB or by the signal recognition particle (SRP).[5-7] The biochemical analysis showed further that the translocon is comprized of a membrane-embedded SecYE channel complex[8] and a peripheral SecA ATPase which functions as a motor to drive translocation (Fig. 1).[9] Genomic analysis revealed that SecYE is highly conserved in Bacteria, Archaea and eukaryotes.[10,11] Isolation of large amounts of SecYE complex or its eukaryotic homolog, the Sec61$\alpha\gamma$ complex, allowed for further biochemical, biophysical and structural analysis. Both complexes copurify with a small subunit, SecG and Sec61β respectively, although these subunits do not share obvious homology.[11] Additional components, such as the heterotrimeric complex SecDFyajC[12] and the proton motive force (PMF),[13] were found to contribute to the Sec pathway but in vitro reconstitution experiments demonstrated that SecA, SecE and SecY are necessary and sufficient for the basal activity

*Corresponding Author: Franck Duong—Department of Biochemistry and Molecular Biology, University of British Columbia, 2146 Health Sciences Mall, Vancouver, B.C., V6T 1Z3, Canada. Email: FDuong@interchange.ubc.ca

Protein Movement Across Membranes, edited by Jerry Eichler. ©2005 Eurekah.com and Springer Science+Business Media.

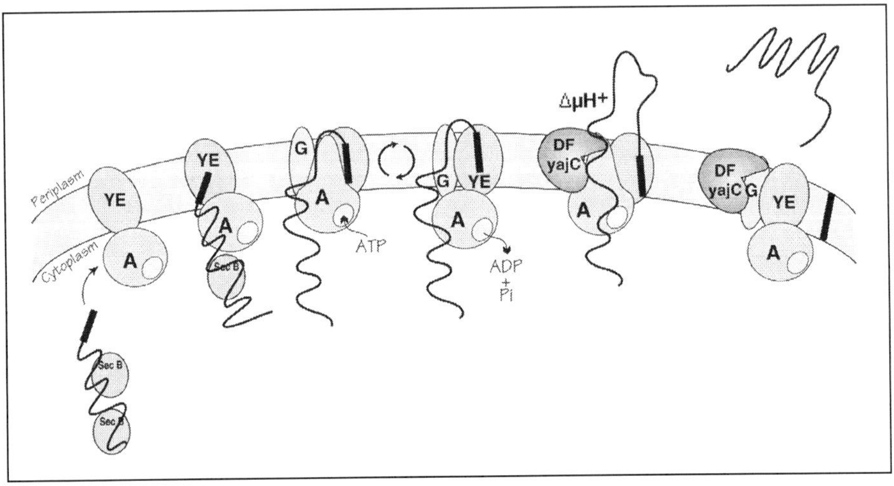

Figure 1. Model for preprotein translocation across the cytoplasmic membrane. See text for details.

of the translocase.[14,15] The following paragraphs describe in finer detail more recent findings and address current questions on the bacterial translocation system.

The SecYEG Translocon at the Atomic Level

Biochemical, biophysical and electrophysiology studies established early on that the Sec complex serves as the channel through which preproteins traverse the membrane. The recent solution of the two-dimensional[16,17] and three-dimensional[18] structures of SecYEG (~75 kDa) and SecYEβ from the bacterium *E. coli* and the archaeon *M. jannaschii*, respectively, provide structural support and new insight into the translocation mechanism. As a full chapter on the Sec channel structure appears elsewhere in this book (see Chapter 4), only a brief description is given here (Fig. 2). The SecY subunit consists of two sub-domains, the transmembrane segments TM1-TM5 and TM6-TM10, arranged like a clamp and related to each other by a two-fold pseudo-symmetry axis. The essential SecE subunit docks its TM helix across the interface of the two SecY domains, clamping them together. The proposed translocation channel is located in the center of the SecY subunit which is filled by a short distorted helix (TM2a, termed the plug) extending halfway to the center of the membrane. Movement of the plug would yield a continuous aqueous channel through which preproteins would be translocated. Halfway across the membrane plane, the channel is also constricted by a ring of six hydrophobic amino acid residues. This ring is proposed to seal the channel but would also widen just enough, probably by shifts in the helices forming the channel, to allow the passage of a polypeptide chain. The small Secβ subunit (SecG-like subunit) is peripherally attached and makes limited contact with SecY, consistent with its nonessential role in translocation. A groove situated between the TM segments at the edges of the two SecY halves (interface between TM2 and TM7) is accessible to the lipid bilayer. As the other sides of SecY are contacted by the SecE and Secβ subunits, this groove may form a lateral gate for release of TM segment of membrane protein.

Binding and Orientation of the Leader Peptide into the Translocon

Leader peptides consist of a short positively charged N-teminal region followed by a central hydrophobic core and a leader peptidase cleavage site. The physicochemical properties

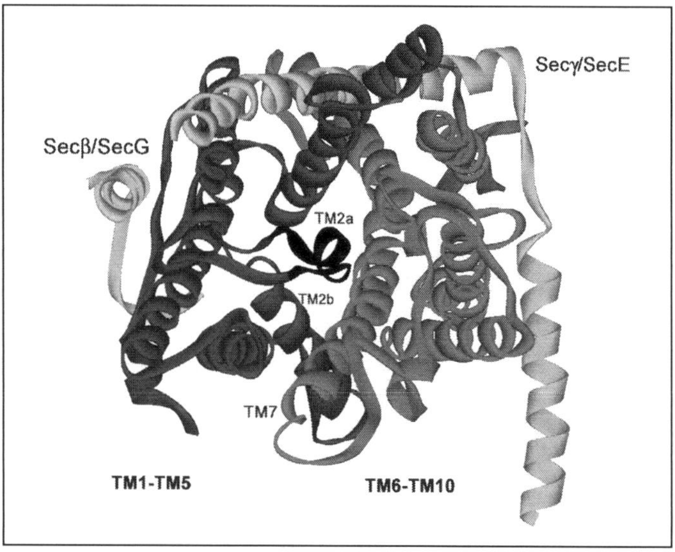

Figure 2. Structure of the SecYEG complex. The structure of the Sec complex from *M. jannaschii* (PDB: 1RHZ) viewed from the periplasmic side. The two halves of Secα/SecY (TM1-TM5 and TM6-TM10) are shown in dark and pale tone respectively. Both halves form a clamp that might open between TM2b and TM7 to deliver transmembrane proteins. TM2a plugs the channel and may move toward SecE and away from the channel cavity, yielding a passage way for a translocating polypeptide.

of leader peptides are essential for correct interaction with the translocon, such that the N-terminus of the leader peptide stays in the cytosol while the hydrophobic core crosses the membrane.[19-21] Insertion of the leader peptide into the channel at an early stage of translocation has been thoroughly analyzed by photo-crosslinking technologies. With photoreactive probes positioned at single sites along the leader peptide, it was shown that the opposite sides of the hydrophobic core contact TM segments 2 and 7 of the SecY subunit.[22-24] Each residue of the leader peptide could also be cross-linked to phospholipids, suggesting that the binding site is located at the interface of the protein channel and the lipid phase.[23,25] These earlier experiments are now supported by the atomic structure of SecY which reveals that TM2 and TM7 are located at the interface of the two SecY halves, adjacent to the pore channel and accessible both from the lipid and cytoplasmic side of the membrane.[18] Moreover, the sequences of TM domains 2 and 7 are well-conserved, suggesting a similar mechanism of leader peptide recognition across evolution.[11] Binding and orientation of the leader peptide into the translocon also involves specific charged residues in the cytoplasmic and periplasmic loops of SecY. Site-directed charge-reversal mutations indicated that these conserved amino-acyl residues functionally interact with charged residues in the N-terminus of the leader peptide in order to set its correct topology in the channel.[26,27] The residues immediately after the signal sequence were found in contact with the SecY subunit but not with lipids, supporting a model in which the polypeptide chain inserts in a loop-like configuration into the channel.[23]

Opening of the Translocation Channel

Pioneering experiments showed that the addition of synthetic leader peptides to the cytoplasmic side of reconstituted *E. coli* membrane bilayer opens aqueous pores detectable by conductivity measurements.[28] The model derived from the atomic structure predicts that two movements may occur in order for the channel to open and to accommodate the leader and attached polypeptide.[18] First, the plug domain may move away from its blocking position in the protein channel and second, the N- and C-terminal halves of SecY may move apart to create a lateral opening of the translocation pore necessary to embrace the polypeptide chain. Alternatively, a diaphragm-like movement of the TM segments would widen the pore sufficiently to allow insertion of the polypeptide chain. Although these putative movements await experimental support, some earlier experiments are in favor of these models. It was shown that unique cysteines introduced in the domain of SecY forming the plug (TM2a) and at the C-terminal end of SecE can form a disulfide bridge.[29] Since these two cysteines are 20Å apart in the closed channel structure, the observed cross-link is now explained by the movement of the plug away from the center of the channel. Moreover, the disulfide bridge formation had a dominant lethal affect,[29] as expected if the channel was locked into a permanently open state by the covalent modification. Other possible experimental support is provided by Prl mutants, a collection of mutations in SecY or SecE which up-regulate the activity of the translocase. Since these mutations allow secretory proteins with defective or even deleted leader peptides to be transported,[30,31] they may mimic the effect of signal sequence binding. A previous study indicated that the Prl mutations increase the conformational flexibility of the translocon,[32] and the atomic structure shows that most of the mutations are located in the center of the channel, particularly on the internal side of TM7 and in the plug.[18] Thus, it is postulated the Prl mutations could increase the dynamics of the plug movement or facilitate widening of the pore during initiation of translocation, and therefore reduce the requirement for a functional leader peptide.

Translocation Pause

Short hydrophobic stretches in the mature domain of preproteins induce a transient pause in the translocation movement which leads to the formation of translocation intermediates across the channel.[33] Deletion or relocation of these hydrophobic segments significantly alters the pattern of intermediates, while increasing the length and hydrophobicity of the stretch can lead to complete translocation arrest.[34,35] These observations suggest that the mechanism involved during translocation pause and translocation arrest are probably similar. The atomic structure shows that the channel is shaped as an hourglass with a constriction of hydrophobic residues in its center. It is proposed that the hydrophobic ring may form a seal around the translocating polypeptide chain while the hourglass shape may serve to limit the contact of the chain with the channel walls.[18] While this organization may minimize the energy required for polypeptide movement through the membrane, it may also serve in the recognition of hydrophobic stretches. If the length and hydrophobicity of the stretch is sufficient to span the membrane, the protein is eventually released into the lipid phase of the membrane.[35,36] The TM segment of a nascent membrane protein has been shown to move from the aqueous interior of the channel to the hydrophobic environment of the lipid bilayer.[25,37] The atomic structure shows that the TM2/TM7 interface is the only possible escape for lateral release of TM segments of membrane proteins in transit through the translocon.

The Quaternary Structure of the SecYEG Translocon

The understanding of the translocation channel is further complicated by the fact that SecYEG exists as dimeric and tetrameric assemblies. Low resolution electron microscopy (EM) images of purified mammalian, yeast and bacterial translocon all revealed the oligomeric state of the Sec complex.[38-41] The stoichiometry of these assemblies was then established using various biochemical and biophysical investigations such as crosslinking,[42] sedimentation analysis[16] and blue-native gel electrophoresis.[43] In the SecYEG dimer, SecE is located at the interface of the protomers, such that the TM2/TM7 interfaces are pointed in opposite directions and toward the lipid bilayer. This is the organization seen in the 2D crystal structure,[17] which is also supported by a cysteine crosslinking study showing that the two SecE subunits are close to one another.[44] The organization of the Sec protomers within the tetrameric assemblies is unknown. It is argued that two dimers, each with an organization similar to that observed in the lattice of the two-dimentional crystals, may associate in a side-by-side manner to form a tetramer.[45] Such organization leaves the lateral gates oriented toward the lipid bilayer but other configurations are, nonetheless, possible. Since evidence for the existence of SecYEG as a monomer in the membrane is lacking, it is unclear why the SecYEG complex exists as an oligomer while a single copy of SecYEG seems to form the translocation channel. The central depression seen in the oligomeric ring-like structure was initially postulated to form a translocation pore.[38-41] In the 2D crystals, the dimeric translocon presents a funnel-like cavity formed by adjacent protomers and closed on its periplasmic face.[17] It is now proposed that the pore-like structure formed by the tetramer or the cavity observed within the dimer in fact reflect a depression at the interface of the protomers rather than a true translocation channel.

Dynamic Behavior of SecYEG Oligomers

It is also unclear whether the Sec complex undergoes transitions in its oligomeric status as part of the translocation event. Native electrophoresis experiments show that SecYEG dimers reversibly dissociate into monomers in a detergent-dependent manner[43] and a protein concentration-dependent equilibrium between Sec tetramers and monomers exists, as detected by analytical centrifugation.[16] However, such dynamic association of Sec protomers takes place in detergent solution and may be different once the SecYEG complex is embedded in the phospholipid bilayer. The reconstitution of membranes containing active SecYEG involves the dilution of detergent and thus prompts the formation of dimers.[43] The same is true for the growth of 2D crystals and only dimeric assemblies formed in the crystallized membrane.[16] Moreover, subunit exchange studies[46] and fluorescence resonance energy transfer experiments[47] failed to observe any exchange of the protomeric components within the membranous oligomer. Altogether, these observations argue that the translocon may not experience rearrangements in its oligomeric status per se, although translocation-related dynamic changes may occur. Indeed, cysteine-scanning mutagenesis identified enhancement of the interhelical SecE contact at the initiation of translocation,[42] suggesting that translocation results in a rearrangement of SecE molecules within the SecYEG oligomer. Electron microscopy analysis of proteoliposome-reconstituted, detergent-solubilized dimeric SecYEG revealed that binding of SecA with nucleotides lead to the recruitment of two SecYEG dimers to form the tetrameric SecYEG assembly.[41] Similarly, reconstitution of the mammalian translocon ring structure required the presence of ribosomes for recruitment of individual Sec61 eukaryotic complexes.[48]

Atomic Structure of the SecA Translocation Motor

The SecA ATPase (~100 kDa) interacts with the SecYEG channel to drive translocation. In addition to its high affinity for SecYEG,[49] SecA also interacts with numerous ligands: leader and mature regions of preproteins, acidic phospholipids, SecB, nucleotides, $Mg,^{2+}$ Zn^{2+} and its own mRNA.[50] Accordingly, the crystal structure of SecA from *B. subtilis* reveals a complex multidomain protein (Fig. 3).[51] The motor ATPase domain is made up of two RecA-like folds (termed nucleotide binding folds, NBF) similar to those found in superfamily 1 and 2 helicases.[52] The interface between NBF1 and NBF2 forms the nucleotide binding site. Three other domains are linked to the ATPase domain: the preprotein cross-link domain (PPXD), helical wing domain (HWD), and helical scaffold domain (HSD).[51] The PPXD domain can be

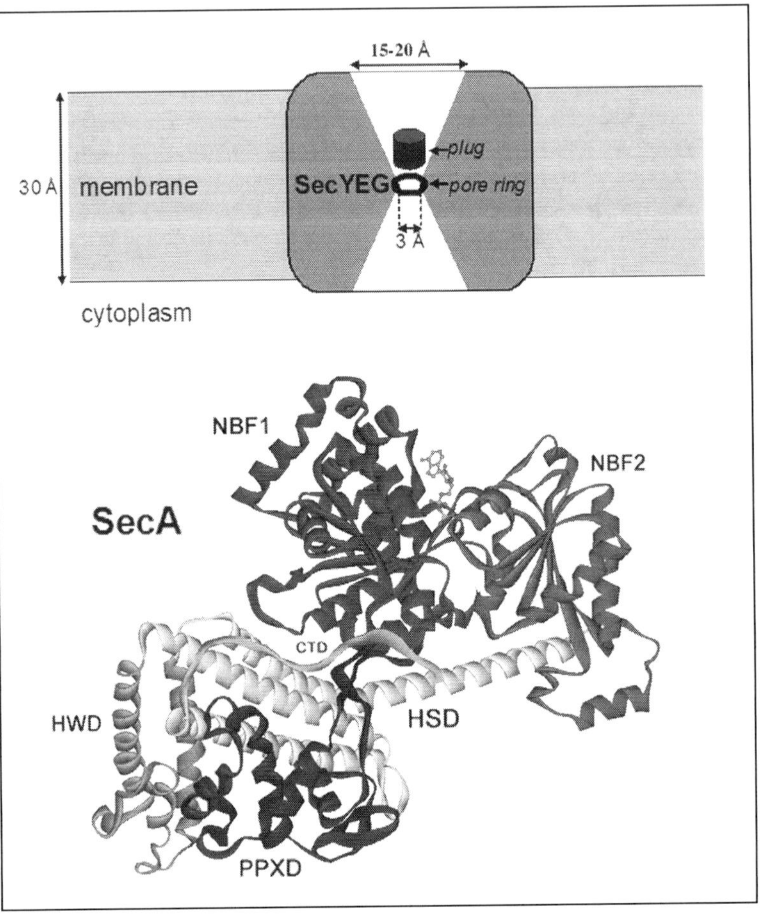

Figure 3. Modular structure of SecA. Ribbon representation of *B. subtilis* SecA (PDB: 1M74). See text for an introduction to NBF1, NBF2, HSD, HWD, PPXD and CTD domains. Above SecA, a fit-to-scale sketch represents the monomeric translocon inside the hydrophobic bilayer of the cytoplasmic membrane. The channel inside the translocon has an hourglass shape with the characteristic dimensions indicated.

cross-linked to the leader and mature domain of preprotein.[53,54] The HSD domain contains a long α-helix acting as a connection between the motor and translocation domains of SecA. The HWD domain is an insertion in the HSD domain and seems to be flexible and loosely linked with the rest of the molecule. The extreme C-terminal region (CTD) is also flexible and has been shown to bind lipid, Zn^{2+} and SecB.[55,56] Recently, comparison of the atomic structures of monomeric and dimeric SecA revealed that SecA monomerization generates relative movement of the PPXD, HSD and HWD domains such that a potential preprotein-binding groove forms at the surface of SecA[57] (see below also).

Binding of the SecA Motor to the SecYEG Channel

The regions of interaction between SecA and SecYEG remain to be characterized. Ligand affinity blotting experiments indicate that SecA binds to the first 107 amino acid residues of SecY[58] and intergenic suppressor studies suggest that the C-terminal cytoplasmic loop (between TM8 and TM9) and the C-terminal tail of SecY are important for SecA interaction.[59-61] Mutations in the C-terminal tail do not abolish the binding of SecA, but prevents its activation.[59] Similarly, random targeted mutagenesis identified residues in the cytoplasmic loop TM8/TM9 as indispensable for productive SecA-dependent translocation, but their effect on SecA binding is unknown.[60] The sequences of SecA that interact with SecYEG are also poorly defined. Extragenic suppressors of SecY mutations map all over SecA molecules.[62] Ligand affinity experiments identified the C-terminal third of SecA as the SecYEG-interacting domain.[58] In contrast, characterization of the binding constant of truncated SecA derivatives indicate that the N-terminal domain comprizing the ATPase motor is responsible for the interaction with SecYEG[63] but both domains may contribute to optimal binding of SecA to the SecYEG complex. Moreover, the stoichiometry of the SecA-SecYEG association is largely unknown (see below).

How Does SecA Use ATP to Catalyze Translocation?

The SecYEG-bound SecA ATPase activity is stimulated by a translocation-competent preprotein.[9] This activity, termed SecA translocation ATPase, is responsible for the preprotein translocation reaction. In vitro reconstitution shows that the initiation step requires ATP binding but not its hydrolysis.[64] This initial event leads the leader peptide and attached polypeptide to cross the channel in a loop-like configuration such that it can be processed by a signal peptidase at the periplasmic face of the membrane. Continued translocation then requires ATP hydrolysis which causes cycles of binding and the release of the preprotein from SecA.[64,65] Under appropriate in vitro conditions, it has been shown that translocation is a stepwise proccess, corresponding to translocation steps of 20-30 amino acid residues of the polypeptidic chain.[64,67] The mechanism by which the energy of ATP binding hydrolysis at SecA is converted into the movement of preproteins across the membrane has been related to the SecA transmembrane mobility at SecYEG, called the insertion-deinsertion cycle.[68] This model is based on the observation that SecA becomes protected from added protease under the conditions of active preprotein translocation.[68,69] Several regions of SecA are indeed accessible for chemical modification from the periplasmic side of the membrane.[70-72] The membrane insertion-desinsertion of SecA is regulated by ATP and repeated cycles of these movements has been proposed to drive the stepwise movement of preprotein across the membrane.[68] Whether SecA truly inserts into the translocon and across the membrane, and how the ATP-derived energy is coupled to the preprotein movement, remains controversial and not yet fully understood. What is clear is that SecA undergoes conformational changes that are coupled to its interaction with ligands and driven by the ATPase cycle. By analogy to the helicase working mechanism,[52] the two RecA-like domains of SecA may move relative to one another during the ATPase cycle, creating domain movements which may be propagated via the long α-helix (HSD) to the other SecA domains to generate preprotein motion.[73,74] Steady-state tryptophan fluorescence anisotropy spectroscopy suggests that nucleotide-free SecA is in a domain-dissociated conformation which may have high affinity for SecYEG.[51] In contrast, nucleotide binding would result in the presentation of compact conformations with low affinity. Thus, both SecA conformational changes and variation in SecYEG-affinity would provide the driving force for translocation to occur.

The SecA Monomer-Dimer Equilibrium

An understanding of the SecA mechanism seems further complicated by the fact that SecA exists in solution as a dimer in equilibrium with a small fraction of monomers.[75,76] The dimeric organization maximizes the buried solvent-accessible surface area and intermolecular protomeric contacts, but the interface between the SecA dimer is not extensive.[51,57] The equilibrium can be shifted towards the monomeric state by relatively small changes in the SecA primary sequence or incubation conditions.[75,77] An early study based on fluorescence resonance energy transfer experiments and involving heterodimers with one ATPase-inactive subunit has suggested that the SecA dimer is the active species in translocation,[78] but this view has been recently challenged. Acidic lipids, which are essential for SecA activation were found to induce dissociation of the dimer.[77] Furthermore, synthetic signal peptides can induce monomerization of SecA while a mutant that fails to dimerize retains some translocation activity.[77,79] Finally, the rotation of the PPXD domain, which generates a large groove probably involved in preprotein binding, occurs upon monomerization of SecA.[57] Altogether, these recent observations suggest that the active form of SecA may be monomeric or that SecA monomerization may be critical at some stage of the translocation reaction. The oligomeric state of SecA during translocation and when bound at SecYEG remains, however, to be determined. Recent native electrophoresis and analytical centrifugation experiments suggest that detergent-solubilized and stabilized dimeric SecYEG can bind both monomeric and dimeric SecA with a stoichiometry modulated by nucleotides.[80,81] The variability in the stoichiometry of the SecYEG-SecA complex might carry significant implications. It is now hypothesized that the mechanism by which SecA mediates protein translocation may resemble the mechanism by which helicases mediate unwinding of nucleic acid duplexes.[82,83] Two distinct mechanisms, called the inch-worm model and the active rolling model, can be envisioned. According to the 'inchworm' mechanism, the SecA monomer is the functionally active species: cycles of ATP binding and hydrolysis would trigger localized conformational changes in the SecYEG-bound SecA monomer, leading to processive feeding of the polypeptide through the channel. According to the 'rolling' model, ATP-driven cycles of SecA monomerization-dimerization would mediate the processive passage of preprotein: a free SecA monomer may bind a new segment of preproteins before reassociating with the SecYEG-bound SecA monomer. In both models, a SecYEG-bound SecA monomer would be maintained in close association with the channel throughout the translocation process.

The Translocase Makes Use of the Proton Motive Force

Preprotein translocation is strongly stimulated by the PMF, both with native membranes and with purified and reconstituted SecYEG translocase.[84,85] Several subreactions of the translocation process seem to be simultaneously affected by the PMF. Earlier studies have shown that PMF can drive forward movement of preprotein translocation intermediates when SecA is no longer associated with the polypeptidic chain.[67,86] The $\Delta\psi$ and ΔpH components of the PMF may act on the preprotein itself via some sort of electrophoretic or folding effect on the polypeptide chain in transit.[87] The binding of leader peptide to the cytoplasmic membrane and its subsequent insertion in the translocation channel may also be optimized by the PMF.[89] Alternatively, or in addition, the PMF may directly modify the conformation of the translocation channel and its subsequent interaction with the translocation partners. Indeed, the Prl mutations which may alter the conformation of the channel render the in vivo and in vitro translocation less PMF-dependent, suggesting that Prl mutations may mimic the effect of the PMF.[90] Furthermore, the PMF accelerates the conformational changes of SecA that occur during translocation[91] and the stimulatory effect of the PMF is more obvious at low SecA concentrations.[92] The same Prl mutations, which decrease the PMF-dependency of translocation,[90] also increase the affinity of SecA for SecYEG.[93] It is thus possible that the PMF could change the conformation of the channel such that it modifies the dynamics of the SecYEG-SecA association.

Additional Subunits Make the Translocase Holo-Enzyme

In contrast to SecA, SecY and SecE, the SecG subunit is not essential for cell viability and translocation and is not conserved outside the bacterial kingdom.[11] SecG is a 12-kDa protein with two TM segments connected via a short apolar cytosolic segment.[94,95] This small subunit enhances the translocation rate and this enhancing effect is particularly seen in vitro. SecG is not needed for the high-affinity binding of SecA to SecYE but it readily stimulates SecA activity.[13,96] The in vivo contribution of SecG is clearly observed only when SecA function is compromised by mutations or when SecA activity may become more critical for translocation such as at low temperatures, in the absence of SecDF, in absence of acidic phospholipids, or at low transmembrane PMF.[13,97] It has been shown that SecG exists in two inverted topological states in the membrane and interconversion between these states is linked to the SecA membrane insertion reaction.[98] SecG may enhance translocation and SecA activity by acting on the conformation of the translocation channel. However, the atomic structure shows that SecG is located at the periphery of the SecYE complex[18] and EM pictures apparently indicate that SecG is not required for the formation of the SecYE ring-like structures.[40]

The core SecYE also associates with the SecDFyajC heterotrimeric membrane protein complex.[13,99] SecD and SecF present sequence similarity, each spanning the membrane six times and possesing a large periplasmic loop between the first and second transmembrane segments.[100] In *B. subtilis*, SecD and SecF are even fused into one large polypeptide.[101] YajC, a small single transmembrane protein, exists in tight complex with SecDF[13,99] and its gene is located in the same operon. Altogether, these observations suggest that the role of these three proteins is somehow linked but their true function remains largely unknown. In vivo, the absence of SecDF, but not YajC, severely affects cell viability and the efficiency of protein translocation.[12] In vitro, the stimulatory effect of SecDFyajC is obvious only when membranes are depleted for SecG, suggesting that the stimulatory function of SecG covers that of SecDFyajC in the reconstituted system.[15] Interestingly, the level of SecG in membranes is decreased upon SecDFyajC depletion and recovered to a normal level when SecDFyajC is expressed,[102] suggesting that a coordinated balance between these stimulatory subunits exist. At low translocation rates, it has been shown SecDFyajC increases the formation and accumulation of preprotein translocation intermediates in transit across the channel.[103] These translocation intermediates are then propelled forward after energization of the membrane by the PMF. This result suggests that SecDFyajC may serve to coordinate the action of ATP- and PMF-driven translocation.[103] SecDFyajC has also been shown to modulate the behaviour of SecA toward stabilization of the membrane-inserted conformation.[104] Since Archaea contain SecD and SecF homologues while a SecA homologue is absent,[101] this later effect may be indirect and rather caused by the stabilization of the translocation intermediate.[103] Finally, it has been proposed that SecD plays a role in protein release following the translocation event.[105]

Concluding Remarks

Genetic and biochemical studies have provided the first elementary and essential insight into the mechanism of preprotein translocation. Structural studies have resulted in a significant advance of our understanding and allowed further interpretation of previously obtained experimental data. As discussed in several places in this chapter, many aspects of the translocation reaction are accompanied by dynamic conformational changes and transient associations. The combination of advanced biochemical tools together with high-resolution structural approaches will soon allow for an exact description of the relation of structure to function, leading to detailed knowledge about the mechanism of protein translocation across and into the cytoplasmic membrane. Finally, further challenges include the description of translocon function integrated in the larger context of the cell physiology, and its cross-talk with the other translocation systems present in the cell envelope.

Acknowledgments

We thank Dr. J. Brunstein for critical reading of the manuscript. FD greatfully acknowledges research support from the Canada Foundation for Innovation, the Canada Research Chair program and the University of British Columbia. APM is supported by a scholarship from INSERM-CIHR.

References

1. Danese PN, Silhavy TJ. Targeting and assembly of periplasmic and outer-membrane proteins in Escherichia coli. Annu Rev Genet 1998; 32:59-94.
2. Bieker KL, Phillips GJ, Silhavy TJ. The sec and prl genes of Escherichia coli. J Bioenerg Biomembr 1990; 22:291-310.
3. Schatz PJ, Beckwith J. Genetic analysis of protein export in Escherichia coli. Annu Rev Genet 1990; 24:215-248.
4. Wickner W, Leonard MR. Escherichia coli preprotein translocase. J Biol Chem 1996; 271:29514-29516.
5. Valent QA, Scotti PA, High S et al. The Escherichia coli SRP and SecB targeting pathways converge at the translocon. EMBO J 1998; 17:2504-2512.
6. Koch HG, Hengelage T, Neumann-Haefelin C et al. In vitro studies with purified components reveal signal recognition particle (SRP) and SecA/SecB as constituents of two independent protein-targeting pathways of Escherichia coli. Mol Biol Cell 1999; 10:2163-2173.
7. Driessen AJ, Manting EH, van der Does C. The structural basis of protein targeting and translocation in bacteria. Nat Struct Biol 2001; 8:492-498.
8. Brundage, L, Fimmel, CJ, Mizushima S et al. SecY, SecE and band 1 form the membrane-embedded domain of Escherichia coli preprotein translocase. J Biol Chem 1992; 267:4166-4170.
9. Lill R, Cunningham K, Brundage LA et al. SecA protein hydrolyzes ATP and is an essential component of the protein translocation ATPase of Escherichia coli. EMBO J 1989; 8:961-966.
10. Hartmann E, Sommer T, Prehn S et al. Evolutionary conservation of components of the protein translocation complex. Nature 1994; 367:654-657.
11. Cao TB, Saier MH. The general protein secretory pathway: Phylogenetic analyses leading to evolutionary conclusions. Biochem Biophys Acta 2003; 1609:115-125.
12. Pogliano JA, Beckwith J. SecD and SecF facilitate protein export in Escherichia coli. EMBO J 1994; 13:554-561.
13. Bakker EP, Randall LL. The requirement for energy during export of beta-lactamase in Escherichia coli is fulfilled by the total protonmotive force. EMBO J 1984; 3:895-900.
14. Brundage L, Henndrick JP, Schiebel E et al. The purified E. coli integral membrane protein SecY/E is sufficient for reconstitution of SecA-dependent precursor protein translocation. Cell 1990; 62:649-657.
15. Duong F, Wickner W. Distinct catalytic roles of the SecYE, SecG and SecDFyajC subunits of preprotein translocase holoenzyme. EMBO J 1997; 16:2756-2768.
16. Collinson I, Breyton C, Duong F et al. Projection structure and oligomeric properties of a bacterial core protein translocase. EMBO J 2001; 20:2462-2471.
17. Breyton C, Haase W, Rapoport TA et al. Three-dimensional structure of the bacterial protein-translocation complex SecYEG. Nature 2002; 418:662-665.
18. Van den Berg B, Clemons Jr WM, Collinson I et al. X-ray structure of a protein-conducting channel. Nature 2004; 427:36-44.
19. Pugsley AP. Translocation of proteins with signal sequences across membranes. Curr Opin Cell Biol 1990; 2:609-616.
20. von Heijne G. Life and death of a signal peptide. Nature 1998; 396:111-113.
21. Martoglio B, Dobberstein B. Signal sequences: More than just greasy peptides. Trends Cell Biol 1998; 8:410-415.
22. Mothes W, Prehn S, Rapoport TA. Systematic probing of the environment of a translocating secretory protein during translocation through the ER membrane. EMBO J 1994; 13:3973-3982.
23. Plath K, Mothes W, Wilkinson BM et al. Signal sequence recognition in posttranslational protein transport across the yeast ER membrane. Cell 1998; 94:795-807.

24. Plath K, Wilkinson BM, Stirling CJ et al. Interactions between Sec complex and prepro-alpha-factor during posttranslational protein transport into the endoplasmic reticulum. Mol Biol Cell 2004; 15:1-10.
25. Martoglio B, Hofmann MW, Brunner J et al. The protein-conducting channel in the membrane of the endoplasmic reticulum is open laterally toward the lipid bilayer. Cell 1995; 81:207-214.
26. Goder V, Junne T, Spiess M. Sec61p contributes to signal sequence orientation according to the positive-inside rule. Mol Biol Cell 2004; 15:1470-1478.
27. Goder V, Spiess M. Molecular mechanism of signal sequence orientation in the endoplasmic reticulum. EMBO J 2003; 22:3645-3653.
28. Simon SM, Blobel G. Signal peptides open protein-conducting channels in E coli. Cell 1992; 69:677-684.
29. Harris CR, Silhavy TJ. Mapping an interface of SecY (PrlA) and SecE (PrlG) by using synthetic phenotypes and in vivo cross-linking. J Bacteriol 1999; 181:3438-3444.
30. Flower AM, Doebele RC, Silhavy TJ. PrlA and PrlG suppressors reduce the requirement for signal sequence recognition. J Bacteriol 1994; 176:5607-5614.
31. Derman AI, Puziss JW, Bassford Jr PJ et al. A signal sequence is not required for protein export in prlA mutants of Escherichia coli. EMBO J 1993; 12:879-888.
32. Duong F, Wickner W. The PrlA and PrlG phenotypes are caused by a loosened association among the translocase SecYEG subunits. EMBO J 1999; 18:3263-3270.
33. Sato K, Mori H, Yoshida M et al. Short hydrophobic segments in the mature domain of ProOmpA determine its stepwise movement during translocation across the cytoplasmic membrane of Escherichia coli. J Biol Chem 1997; 272:5880-5886.
34. Sato K, Mori H, Yoshida M et al. In vitro analysis of the stop-transfer process during translocation across the cytoplasmic membrane of Escherichia coli. J Biol Chem 1997; 272:20082-20087.
35. Duong F, Wickner W. Sec-dependent membrane protein biogenesis: SecYEG, preprotein hydrophobicity and translocation kinetics control the stop-transfer function. EMBO J 1998; 17:696-705.
36. Saaf A, Wallin E, von Heijne G. Stop-transfer function of pseudo-random amino acid segments during translocation across prokaryotic and eukaryotic membranes. Eur J Biochem 1998; 251:821-829.
37. Heinrich SU, Mothes W, Brunner J et al. The Sec61p complex mediates the integration of a membrane protein by allowing lipid partitioning of the transmembrane domain. Cell 2000; 102:233-244.
38. Beckmann R, Bubeck D, Grassucci R et al. Alignment of conduits for the nascent polypeptide chain in the ribosome-Sec61 complex. Science 1997; 278:2123-2126.
39. Ménétret JF, Neuhof A, Morgan DG et al. The structure of ribosome-channel complexes engaged in protein translocation. Mol Cell 2000; 6:1219-1232.
40. Meyer TH, Ménétret JF, Breitling R et al. The bacterial SecY/E translocation complex forms channel-like structures similar to those of the eukaryotic Sec61p Complex. J Mol Biol 1999; 285:1789-1800.
41. Manting EH, van Der Does C, Remigy H et al. SecYEG assembles into a tetramer to form the active protein translocation channel. EMBO J 2000; 19:852-861.
42. Kaufmann A, Manting EH, Veenendaal AK et al. Cysteine-directed cross-linking demonstrates that helix 3 of SecE is close to helix 2 of SecY and helix 3 of a neighboring secE. Biochemistry 1999; 38:9115-9125.
43. Bessonneau P, Besson V, Collinson I et al. The SecYEG preprotein translocation channel is a conformationally dynamic and dimeric structure. EMBO J 2002; 21:995-1003.
44. Veenendaal AK, Van Der Does C, Driessen AJ. The core of the bacterial translocase harbors a tilted transmembrane segment 3 of SecE. J Biol Chem 2002; 277:36640-36645.
45. Clemons Jr WM, Menetret JF, Akey CW et al. Structural insight into the protein translocation channel. Curr Opin Struct Biol 2004; 14:390-396.
46. Yahr TL, Wickner WT. Evaluating the oligomeric state of SecYEG in preprotein translocase. EMBO J 2000; 19:4393-4401.
47. Mori H, Tsukazaki T, Masui R et al. Fluorescence resonance energy transfer analysis of protein translocase. SecYE from Thermus thermophilus HB8 forms a constitutive oligomer in membranes. J Biol Chem 2003; 278:14257-14264.

48. Hanein D, Matlack KE, Jungnickel B et al. Oligomeric rings of the Sec61p complex induced by ligands required for protein translocation. Cell 1996; 87:721-732.
49. Hartl FU, Lecker S, Schiebel E et al. The binding cascade of SecB to SecA to SecY/E mediates preprotein targeting to the E. coli plasma membrane. Cell 1990; 63:269-279.
50. Economou A. Bacterial secretome: The assembly manual and operating instructions. Mol Membr Biol 2002; 19:159-169.
51. Hunt JF, Weinkauf S, Henry L et al. Nucleotide control of interdomain interactions in the conformational reaction cycle of SecA. Science 2002; 297:2018-2026.
52. Caruthers JM, McKay DB. Helicase structure and mechanism. Curr Opin Struct Biol 2002; 12:123-133.
53. Kimura E, Akita M, Matsuyama S et al. Determination of a region in SecA that interacts with presecretory proteins in Escherichia coli. J Biol Chem 1991; 266:6600-6606.
54. Baud C, Karamanou S, Sianidis G et al. Allosteric communication between signal peptides and the SecA protein DEAD motor ATPase domain. J Biol Chem 2002; 277:13724-13731.
55. Breukink E, Nouwen N, van Raalte A et al. The C terminus of SecA is involved in both lipid binding and SecB binding. J Biol Chem 1995; 270:7902-7907.
56. Fekkes P, de Wit JG, Boorsma A et al. Zinc stabilizes the SecB binding site of SecA. Biochemistry 1999; 38:5111-5116.
57. Osborne AR, Clemons Jr WM, Rapoport TA. A large conformational change of the translocation ATPase SecA. Proc Natl Acad Sci USA 2004; 101:10937-10942.
58. Snyders S, Ramamurthy V, Oliver D. Identification of a region of interaction between Escherichia coli SecA and SecY proteins. J Biol Chem 1997; 272:11302-11306.
59. Matsumoto G, Yoshihisa T, Ito K. SecY and SecA interact to allow SecA insertion and protein translocation across the Escherichia coli plasma membrane. EMBO J 1997; 16:6384-6393.
60. Mori H, Ito K. An essential amino acid residue in the protein translocation channel revealed by targeted random mutagenesis of SecY. Proc Natl Acad Sci USA 2001; 98:5128-5133.
61. Chiba K, Mori H, Ito K. Roles of the C-terminal end of SecY in protein translocation and viability of Escherichia coli. J Bacteriol 2002; 184:2243-2250.
62. Matsumoto G, Nakatogawa H, Mori H et al. Genetic dissection of SecA: Suppressor mutations against the secY205 translocase defect. Genes Cells 2000; 5:991-999.
63. Dapic V, Oliver D. Distinct membrane binding properties of N- and C-terminal domains of Escherichia coli SecA ATPase. J Biol Chem 2000; 275:25000-25007.
64. Schiebel E, Driessen AJM, Hartl FU et al. $\Delta\mu H+$ and ATP function at different steps of the catalytic cycle of preprotein translocase. Cell 1991; 64:927-939.
65. de Keyzer J, van der Does C, Kloosterman TG et al. Direct demonstration of ATP-dependent release of SecA from a translocating preprotein by surface plasmon resonance. J Biol Chem 2003; 278:29581-29586.
66. Uchida K, Mori H, Mizushima S. Stepwise movement of preproteins in the process of translocation across the cytoplasmic membrane of Escherichia coli. J Biol Chem 1995; 270:30862-30868.
67. van der Wolk JP, de Wit JG, Driessen AJM. The catalytic cycle of the Escherichia coli SecA ATPase comprises two distinct preprotein translocation events. EMBO J 1997; 16:7297-7304.
68. Economou A, Wickner W. SecA promotes preprotein translocation by undergoing ATP-driven cycles of membrane insertion and deinsertion. Cell 1994; 78:835-843.
69. Eichler J, Wickner W. Both an N-terminal 65-kDa domain and a C-terminal 30-kDa domain of SecA cycle into the membrane at SecYEG during translocation. Proc Natl Acad Sci USA 1997; 94:5574-5581.
70. Kim YJ, Rajapandi T, Oliver D. SecA protein is exposed to the periplasmic surface of the E. coli inner membrane in its active state. Cell 1994; 78:845-853.
71. van der Does C, den Blaauwen T, de Wit JG et al. SecA is an intrinsic subunit of the Escherichia coli preprotein translocase and exposes its carboxyl terminus to the periplasm. Mol Microbiol 1996; 22:619-629.
72. Ramamurthy V, Oliver D. Topology of the integral membrane form of Escherichia coli SecA protein reveals multiple periplasmically exposed regions and modulation by ATP binding. J Biol Chem 1997; 272:23239-23246.

73. Sianidis G, Karamanou S, Vrontou E et al. Cross-talk between catalytic and regulatory elements in a DEAD motor domain is essential for SecA function. EMBO J 2001; 20:961-970.
74. Fak JJ, Itkin A, Ciobanu DD et al. Nucleotide exchange from the high-affinity ATP-binding site in SecA is the rate-limiting step in the ATPase cycle of the soluble enzyme and occurs through a specialized conformational state. Biochemistry 2004; 43:7307-7327.
75. Woodbury RL, Hardy SJ, Randall LL. Complex behavior in solution of homodimeric SecA. Protein Sci 2002; 11:875-882.
76. Ding H, Hunt JF, Mukerji I et al. Bacillus subtilis SecA ATPase exists as an antiparallel dimer in solution. Biochemistry 2003; 42:8729-8738.
77. Or E, Navon A, Rapoport T. Dissociation of the dimeric SecA ATPase during protein translocation across the bacterial membrane. EMBO J 2002; 21:4470-4479.
78. Driessen AJ. SecA, the peripheral subunit of the Escherichia coli precursor protein translocase, is functional as a dimer. Biochemistry 1993; 32:13190-13197.
79. Benach J, Chou YT, Fak JJ et al. Phospholipid-induced monomerization and signal-peptide-induced oligomerization of SecA. J Biol Chem 2003; 278:3628-3638.
80. Duong F. Binding, activation and dissociation of the dimeric SecA ATPase at the dimeric SecYEG translocase. EMBO J 2003; 22:4375-4384.
81. Tziatzios C, Schubert D, Lotz M et al. The bacterial protein-translocation complex: SecYEG dimers associate with one or two SecA molecules. J Mol Biol 2004; 340:513-524.
82. Soultanas P, Wigley DB. DNA helicases: 'Inching forward'. Curr Opin Struct Biol 2000; 10:124-128.
83. Velankar SS, Soultanas P, Dillingham MS et al. Crystal structures of complexes of PcrA DNA helicase with a DNA substrate indicate an inchworm mechanism. Cell 1999; 97:75-84.
84. Driessen AJM. Precusor protein translocation by the Escherichia coli translocase is directed by the protonmotive force. EMBO J 1992; 11:847-853.
85. Geller BL, Green HM. Translocation of pro-OmpA across inner membrane vesicles of Escherichia coli occurs in two consecutive energetically distinct steps. J Biol Chem 1989; 264:16465-16469.
86. Tani K, Shiozuka K, Tokuda H et al. In vitro analysis of the process of translocation of OmpA across the Escherichia coli cytoplasmic membrane. A translocation intermediate accumulates transiently in the absence of the proton motive force. J Biol Chem 1989; 264:18582-18588.
87. Driessen AJM,Wickner WT. Proton transfer is rate-limiting for translocation of precursor proteins by the Escherichia coli translocase. Proc Natl Acad Sci USA 1991; 88:2471-2475.
88. van Dalen A, Killian A, de Kruijff B. Delta-psi stimulates membrane translocation of the C-terminal part of a signal sequence. J Biol Chem 1999; 274:19913-19918.
89. Daniels CJ, Bole DG, Quay SC et al. Role for membrane potential in the secretion of protein into the periplasm of Escherichia coli. Proc Natl Acad Sci USA 1981; 78:5396-5400.
90. Nouwen N, de Kruijff B, Tommassen J. PrlA suppressors in Escherichia coli relieve the proton electrochemical gradient dependency of translocation of wild-type precursors. Proc Natl Acad Sci USA 1996; 93:5953-5957.
91. Nishiyama KI, Fukuda A, Morita K et al. Membrane deinsertion of SecA underlying proton motive force-dependent stimulation of protein translocation. EMBO J 1999; 18:1049-1058.
92. Yamada H, Matsuyama S, Tokuda H et al. A high concentration of SecA allows proton motive force-independent translocation of a model secretory protein into Escherichia coli membrane vesicles. J Biol Chem 1989; 264:18577-18581.
93. van der Wolk JP, Fekkes P, Boorsma A et al. PrlA4 prevents the rejection of signal sequence defective preproteins by stabilizing the SecA-SecY interaction during the initiation of translocation. EMBO J 1998; 17:3631-3639.
94. Nishiyama K, Mizushima S, Tokuda H. A novel membrane protein involved in protein translocation across the cytoplasmic membrane of Escherichia coli. EMBO J 1993; 12:3409-3415.
95. Nishiyama K, Hanada M, Tokuda H. Disruption of the gene encoding p12 (SecG) reveals the direct involvement and important function of SecG in the protein translocation of Escherichia coli at low temperature. EMBO J 1994; 13:3272-3277.
96. Matsumoto G, Mori H, Ito K. Roles of SecG in ATP- and SecA-dependent protein translocation. Proc Natl Acad Sci USA 1998; 95:13567-13572.
97. Hanada M, Nishiyama K, Tokuda H. SecG plays a critical role in protein translocation in the absence of the proton motive force as well as at low temperature. FEBS Lett 1996; 381:25-28.

98. Nishiyama K, Suzuki T, Tokuda H. Inversion of the membrane topology of SecG coupled with SecA-dependent preprotein translocation. Cell 1996; 85:71-81.
99. Nouwen N, Driessen AJM. SecDFyajC forms a heterotetrameric complex with YidC. Mol Microbiol 2002; 44:1397-1405.
100. Pogliano KJ, Beckwith J. Genetic and molecular characterization of the Escherichia coli secD operon and its products. J Bacteriol 1994; 176:804-814.
101. Bolhuis A, Broekhuizen CP, Sorokin A et al. SecDF of Bacillus subtilis, a molecular Siamese twin required for the efficient secretion of proteins. J Biol Chem 1998; 273:21217-21224.
102. Kato Y, Nishiyama K, Tokuda H. Depletion of SecDF-YajC causes a decrease in the level of SecG: Implication for their functional interaction. FEBS Lett 2003; 550:114-118.
103. Duong F, Wickner W. The SecDFyajC domain of preprotein translocase controls preprotein movement by regulating SecA membrane cycling. EMBO J 1997; 16:4871-4879.
104. Economou A, Pogliano JA, Beckwith J. SecA membrane cycling at SecYEG is driven by distinct ATP binding and hydrolysis events and is regulated by SecD and SecF. Cell 1995; 83:1171-1181.
105. Matsuyama S, Fujita Y, Mizushima S. SecD is involved in the release of translocated secretory proteins from the cytoplasmic membrane of Escherichia coli. EMBO J 1993; 12:265-270.

CHAPTER 3

Protein Translocation in Archaea

Jerry Eichler*

Abstract

While the process of protein translocation has been extensively addressed in Bacteria and Eukarya, little is known of how proteins cross the membranes of Archaea, the third domain of Life. Analysis thus far suggests the hybrid-like nature of the archaeal protein translocation system, combining selected aspects of the bacterial and eukaryal processes together with Archaea-specific features. The archaeal translocation apparatus simultaneously incorporates homologues of system components found either in Bacteria or Eukarya but not in both, yet seemingly does not include other important elements of these two systems. Moreover, certain facets of the archaeal protein translocation process appear specific to this domain, possibly reflecting adaptations to the extreme environments in which Archaea exist.

Introduction

First identified on the basis of their unique 16S ribosomal RNA secondary structure and later delineated by genomic comparisons, Archaea represent a separate branch of the phylogenetic tree that also includes Bacteria and Eukarya.[1,2] While Archaea have been shown to be distributed across a wide range of biological niches,[3] these microorganisms remain best known as extremophiles, able to thrive in extremes of temperature, pH and salinity, as well as other harsh environments.[4] As such, the archaeal plasma membrane must not only retain its structural integrity in the face of drastic physical surroundings, but also perform a variety of biological activities, including nutrient uptake, cell division, bioenergy production and protein secretion. Hence, better understanding of how membrane-based functions are performed in Archaea would not only describe molecular strategies employed by archaeal membranes in response to extreme conditions, but could also provide new insight into the processes themselves.

In Archaea, a variety of proteins must traverse or insert into the plasma membrane. As in Bacteria and Eukarya, translocation of such proteins requires that they be identified as destined to reside beyond the cytosol and then be delivered to membranous translocation sites where they traverse the membrane. However, unlike our relatively advanced understanding of protein translocation in the other two domains of Life,[5-7] little is known of the steps involved in archaeal protein translocation.[8,9] In the following, the current state of understanding of the archaeal protein translocation process is considered.

*Jerry Eichler—Department of Life Sciences, Ben Gurion University, P.O. Box 653, Beersheva, 84105, Israel. Email: jeichler@bgu.ac.il

Archaeal Signal Peptides

Across evolution, proteins destined for export are usually synthesized as precursors, bearing a cleavable N-terminal signal peptide involved in directing such proteins to the membrane-localized translocation machinery.[5-10] As archaeal signal peptides have only been experimentally confimed in a limited number of examples, it remains unclear what Archaea-specific signal peptides look like. Examination of known and predicted archaeal signal peptides reveals similarities to identified signal peptides from both Eukarya and Bacteria. As such, archaeal preproteins heterologously expressed in Bacteria and Eukarya are effectively secreted.[11-14] Furthermore, since archaeal signal peptidase I, the enzyme responsible for signal peptide cleavage, may function in a manner analogous to the eukaryal enzyme[15,16] (see below), it can be argued that archaeal signal peptides more closely resemble their eukaryal counterparts. In contrast, it has been suggested that archaeal signal peptides are more similar to those employed by Gram-positive Bacteria.[17] Alternatively, archaeal signal peptides may incorporate a hybrid of bacterial and eukaryal traits. Bioinformatic analysis of proposed *Methanococcus jannaschii* and *Sulfolobus solfataricus* signal peptides suggests the presence of a Eukarya-like cleavage site together with a Bacteria-like charge distribution, combined with a unique, Archaea-specific hydrophobic region.[18,19] In other species, however, different rules may apply.[20] Moreover, the existence of signal peptides bearing unique archaeal traits cannot be dismissed. Such signal peptides would likely have been overlooked in those earlier studies relying on similarities to identified eukaryal and bacterial signal peptides. Indeed, archaeal flagellin proteins contain uncharacteristic signal peptides.[21] Finally, whereas signal peptides of the Sec system are thought to predominate in Archaea,[20,22,23] studies addressing the halophilic archaea *Halobacterium* sp. NRC-1 have proposed that use of signal peptides recognized by the twin-arginine translocation (Tat) system predominates in this species.[22,24]

Archaeal Protein Translocation: A Co- or Post-Translational Event?

In Archaea, the relation of protein translation to protein translocation is not known. Indeed, evidence supporting both post- and cotranslational translocation systems have been presented.

In considering the biosynthesis of *Halobacterium salinarum* bacterioopsin, the apoprotein form of the multi-membrane-spanning light-driven proton pump bacteriorhodopsin, evidence favoring a cotranslational translocation system has been presented. Original support for the interplay between translation and translocation was provided by experiments showing co-sedimentation of 7S RNA, a component of the signal recognition particle (SRP, see below), and bacterioopsin mRNA with membrane-bound polysomes, as well as puromycin-induced release of 7S RNA from the polysomes.[25] Later in vivo kinetic labeling experiments confirmed the cotranslational insertion of the N-terminal region of the protein, but also revealed the post-translational insertion of the C-terminal region.[26,27] On the other hand, heterologous expression of a chimeric fusion protein including bacterioopsin in *Haloferax volcanii* revealed the need for the seventh and final transmembrane domain for membrane insertion, suggesting bacterioopsin insertion to take place post-translationally.[28]

In experiments aimed at discerning the temporal relation between translation and secretion in Archaea, *H. volcanii* cells were engineered to express chimeric secretory precursors containing the signal peptide of the major secretory protein of this species i.e., the surface layer glycoprotein, fused to different reporter proteins.[29] By following secretion of the chimera either in the absence and presence of an antibiotic inhibitor of archaeal protein synthesis, it was concluded that translation and secretion occur independentally of each other. The ability of Archaea to secrete proteins in a post-translational manner is intriguing, given the apparent absence of an archaeal homologue of SecA, the ATPase that drives post-translational translocation in Bacteria.[8,9] Finally, the archaeal Tat system may also translocate proteins in a post-translational manner.[22]

Table 1. Components of the SRP pathway across evolution

	Eukarya	Bacteria	Archaea
SRP			
	7S RNA	6S RNA (*B. subtilis*)	7S RNA
	sRNA-85 (trypanosomes)	4.5S RNA (*E. coli*)	
	SRP54	SRP54	SRP54
	SRP19	HBsu (*B. subtilis*)	SRP19
	SRP72		
	SRP68		
	SRP14		
	SRP9		
	SRP21 (yeast)		
	SRP43 (chloroplast)		
SRP Receptor			
	SRα	FtsY	FtsY
	SRβ		

The SRP Pathway and Ribosome Binding in Archaea

The coupling of protein translation to protein translocation in Eukarya and Bacteria is mediated by SRP and the SRP receptor.[30,31] In Archaea, SRP is comprised of 2 protein components, i.e., SRP54 and SRP19, together with an SRP RNA that is highly similar to its eukaryal homologue (Table 1).[32] Despite an overall low extent of sequence conservation, archaeal SRP RNA can assume a secondary structure essentially identical to that of folded human SRP RNA, with the archaeal molecule being distinguished by the presence of helix 1, formed upon pairing of the 5' and 3' ends, and the absence of helix 7. [32] Indeed, archaeal SRP54, SRP19 and SRP RNA are organized into a complex strikingly reminiscent of the better-characterized mammalian SRP.

Recent reconstitutions of SRP and SRP sub-complexes from several different strains have allowed for more detailed study of the archaeal particle.[33-38] For instance, as in mammals, archaeal SRP19 interacts with SRP RNA so as to facilitate SRP54 binding to SRP RNA.[31,34] However, in contrast to the eukaryal mode of SRP assembly, interaction between SRP RNA and SRP54 is not entirely SRP19-dependent in Archaea, with significant amounts of binding taking place without SRP19.[33,39] This situation has allowed for an assessment of the role of SRP19 in SRP assembly. It is thought that interaction of SRP19 with SRP RNA helix 6 leads to positional or folding changes in SRP RNA helix 8, leading, in turn, to increased SRP54 binding.[33,34] The ability of SRP RNA and SRP54 to interact in the absence of SRP19 could reflect a need for a more stable SRP in Archaea, possibly related to the extreme environments inhabited by these microorganisms. [34]

Whereas archaeal SRP is reminiscent of its eukaryal counterpart, the archaeal SRP receptor is more similar to FtsY, the bacterial SRP receptor. Like its bacterial homologue, archaeal FtsY exists in both a soluble and membrane-associated form.[40-42] Given the apparent absence of a membranous FtsY receptor in Archaea (as in Bacteria), the manner by which FtsY interacts with the membrane remains, however, unclear. In *Escherichia coli*, membrane binding of FtsY is thought to be mediated via clusters of lysine and arginine residues situated close to the N-terminus of the protein.[43-45] Examination of FtsY sequences from a variety of archaeal species has also revealed the presence of clusters of positively-charged residues at the beginning of the protein.[42] In the case of halophilic Archaea, however, far fewer of such residues are detected, suggesting that archaeal FtsY may rely on additional portions of the protein for membrane association. Indeed, it has been shown that the C-terminal NG domain of *H. volcanii* FtsY is capable of membrane binding.[42]

At some point during the SRP protein targeting cycle, SRP interacts with its receptor, irrespective of the mode of receptor interaction with the membrane. In the eukaryal system, the membrane-localized receptor binds a tertiary ribosome-nascent polypeptide-SRP complex. In Bacteria, the order of events leading to the interaction between SRP and FtsY remains an open question.[30] Similarly, the interplay between archaeal SRP and FtsY is poorly defined. In the hyperthermoacidophilic archaea *Acidianus ambivalens*, the formation of a soluble SRP54-FtsY complex, as well as the membrane binding ability of both SRP54 and FtsY, were reported.[39] In contrast, the membrane binding of *H. volcanii* SRP54 was shown to be FtsY-dependent.[42] In light of this apparent discrepancy, additional study of the interactions that occur within the archaeal SRP pathway are clearly called for. Moreover, it should also be noted that the in vitro membrane-associating behavior of archaeal SRP54 may not reflect the in vivo behavior of the intact ribonucleoprotein particle.

Regardless of the order of binding events, the SRP pathway is responsible for targeting selected translating ribosomes to the membrane. In eukaryal cotranslational protein translocation, SRP delivers a subset of translating ribosomes to the ER membrane through the affinities of SRP for its membranous receptor[46,47] and of the ribosome for the Sec61p-based translocon.[48,49] While understanding the behavior of a bacterial cotranslational translocation pathway is currently the focus of substantial efforts, bacterial ribosomes have also been shown to specifically bind to SecYEG complexes.[50,51] In recent in vitro studies, the ability of functional ribosomes to bind to SecYE-based sites in the haloarchaea *H. volcanii* was shown, confirming the binding of ribosomes to the translocon in all three domains of Life.[52] Indeed, the affinity of archaeal ribosome binding was similar to that measured in Eukaya and Bacteria. Moreover, the non-translating bound haloarchaeal ribosomes remained membrane-associated even following washes with solutions containing up to 3 M KCl. This is in striking difference to the binding profile of eukaryal and bacterial ribosomes, where bound ribosomes are readily released by low salt levels,[46,50,53] and likely reflects the highly saline nature of the cytoplasm in halophilic archaea.[54,55]

The Archaeal Translocon and Other Auxilliary Proteins

In Eukarya, secretory and membrane proteins cross the ER membrane at the translocon, a membrane protein complex based on Sec61$\alpha\beta\gamma$ proteins,[56] while in Bacteria, translocation transpires at the homologous SecYEG complex.[57] While these proteins form the core of the translocon, additional proteins may participate in the translocation event. Analysis of completed archaeal genomes together with the isolation of genes encoding translocon and related components from other species reveals an archaeal translocation apparatus that can be best described as a mosaic of the eukaryal and bacterial complexes (Fig. 1).

Like other SecY/Sec61α proteins, archaeal SecY proteins cross the membrane 10 times[58-62] and thus likely comprise the protein-conducting channel of the translocon. Although named after the bacterial homologue largely due to historical reasons, reported archaeal SecY sequences are far more reminiscent of eukaryal Sec61α proteins than of bacterial SecY.[10,60,63] Similarly, phylogenetic comparisons reveal that archaeal SecE is closer to the eukaryal version of the protein i.e., Sec61γ.[60] Indeed, relying on the similarity of the *S. solfataricus secE* gene with eukaryal Sec61γ sequences, together with the similar positions of archaeal and bacterial *secE* genes within their operons, the homology between bacterial SecE and eukaryal Sec61γ was uncovered.[64]

Along with the core SecYE complex, the bacterial translocon also includes SecG,[65-67] while in Eukarya, Sec61β exists in complex together with Sec61$\alpha\gamma$.[56] Unlike the clear similarities between SecYE and Sec61$\alpha\gamma$ polypeptides, respectively, SecG and Sec61β are seemingly not homologous[60,68] and as such, apparently fulfill distinct translocation functions.[69-72] In

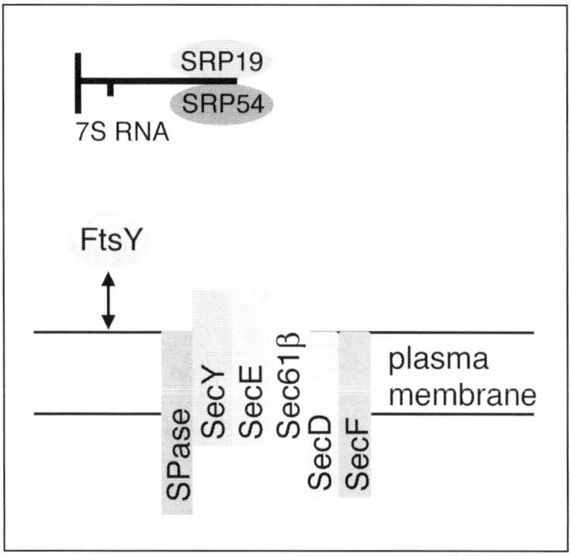

Figure 1. Known components of the Sec translocation pathway in Archaea. A schematic portrayal of those components of the Sec pathway identified either through complete genomic analysis, isolation of individual genes, or via protein expression or purification.

Archaea, neither SecG nor Sec61β were originally detected, although more advanced PSI-BLAST searches later identified sequences corresponding to an archaeal version of Sec61β.[73]

Apart from sequence comparisons, however, little is known of archaeal SecYEβ at the protein level. Complementation of a temperature-sensitive *secY E. coli* mutant with the *Methanococcus vanielii* SecY-encoding gene allowed for growth at non-permissive temperatures,[58] reflecting not only the ability of an archaeal SecY to functionally replace its bacterial counterpart, but also that the archaeal protein is active in a lipid environment different from the ether-based phospholipids that comprise the archaeal membrane.[74] In the haloarchaea *H. volcanii*, chimeric proteins containing SecYE proteins have been isolated and shown to be stably expressed and exclusively localized to the plasma membrane.[61] Most recently, the 3D crystal structure of the *M. jannaschii* SecYEβ complex was solved, offering the first high-resolution glimpse at the structure of the translocon.[62] For a detailed description of the structure of the archaeal translocon, the reader is directed to the chapter by Collinson in this volume (see Chapter 4).

In Bacteria, SecYEG can be found as part of a larger complex that includes SecDF.[71] Some, but not all, completed archaeal genomes contain SecDF homologues, as is also the case in Bacteria.[75,76] A comparison of SecDF sequences reveals that the bacterial and archaeal versions of each protein present similar membrane topologies and positioning of conserved sequence elements, although the makeup of these conserved motifs can be sharply divided along Archaea-Bacteria lines.[76] While the role(s) of SecDF in protein translocation have yet to be clearly defined, these components have been reported to modulate the in vitro membrane-associating behavior of SecA, the ATPase component of the bacterial protein translocation apparatus.[77,78] Hence, given the apparent absence of SecA in Archaea,[8,9] the distinctiveness of conserved sequence elements in archaeal and bacterial SecDF could reflect differences in the functions of these proteins in each domain.

Along with SecDF, SecYEG can be co-isolated with YidC,[79] a member of a protein family involved in the insertion of bacterial, mitochondrial and thylakoid membrane proteins.[80-82]

YidC has also been shown to catalyze Sec-independent membrane protein insertion.[83] While studies have proposed the existence of members of the YidC/Oxa1/Alb3 protein family in Archaea,[84,85] the extent of similarity of the archaeal proteins to known family members is relatively low. The putative archaeal YidC proteins are generally smaller than their bacterial counterparts yet display a similar topology, as has been reported to be the general case for archaeal transport proteins.[86]

The eukaryal translocon can also be found in association with other translocation-related components, including the translocon-associated protein complex[88] and the translocating chain-associated membrane protein.[89] To date, no archaeal homologues of these proteins have been identified.

Sec-Independent Protein Translocation in Archaea

Some, but not all, Archaea also encode for homologues of TatA/E, TatB and TatC,[10,23,87] components of the Sec-independent twin-arginine translocation (Tat) pathway, employed for the translocation of folded proteins, often in complex with co-factors.[90,91] Specifically, sequenced genomes from the archaeal sub-division *Crenarchaeota* all contain genes encoding these components, whereas the same genes are detected in only some genomes of *Euryarchaeota*, the second major sub-division of Archaea.[23] Moveover, various strains are predicted to express differing numbers of Tat pathway components. Genome-wide searches for substrates bearing Tat system signal peptides, readily identified and distinguished from Sec system signal peptides by the presence of a twin arginine-based motif, amongst other traits, also predicts differing degrees of usage of this system in Archaea.[22,23,24,87] For example, such analyses predict the Tat pathway to be the predominant protein translocation system of halophilic archaea.[22,24] It has been argued that extensive use of the Tat system by haloarchaea would allow secretory proteins to first assume their final tertiary structures in the highly saline cytosol of such species and only then traverse the plasma membrane, thereby avoiding potential folding mishaps. On the other hand, genomic analysis of *Methanopyrus kandleri* AV19 predicts an apparent absence of Tat system substrates, despite the presence of TatA/E-encoding genes.[23]

Signal Peptide Cleavage in Archaea

At some later stage in the protein translocation event, the signal peptide that served to target a protein for translocation is removed by the actions of type I signal peptidase.[92,93] While sequence alignment reveals that all type I signal peptidases contain five regions of sequence homology, termed boxes A-E,[15,93,94] the catalytic mechanism and oligomeric status of the enzyme has not been maintained across evolution (Fig. 2). In Bacteria, enzymatic activity relies on a catalytic dyad comprising Box B Ser90 and Box D Lys145 (*E. coli* numbering) residues. In the eukaryal type I signal peptidase, the strictly conserved lysine is replaced by a histidine residue.[15,93,94] Indeed, site-directed mutagenesis studies argue against a role for lysine residues in the catalytic mechanism of the eukaryl enzyme.[95] Thus, the catalytic mechanism of the eukaryal version of signal peptidase remains unknown. Differences in catalytic mechanism are also reflected in the unique pharmacological profiles of bacterial and eukaryal signal peptidases.[92] The bacterial and eukaryal enzymes can further be distinguished by the fact that whereas bacterial signal peptide functions as a single polypeptide, the eukaryal enzyme exists as part of a multi-subunit complex.[96]

Signal peptide cleavage in Archaea represents yet another example of the mosaic nature of archaeal protein translocation discussed above, with the archaeal signal peptidase combining various bacterial, eukaryal and archaeal traits. For example, while the archaeal enzyme has replaced the conserved lysine of the bacterial serine-lysine catalytic dyad with a histidine residue, as in Eukarya, the archaeal signal peptidase appears to function independently, like its bacterial

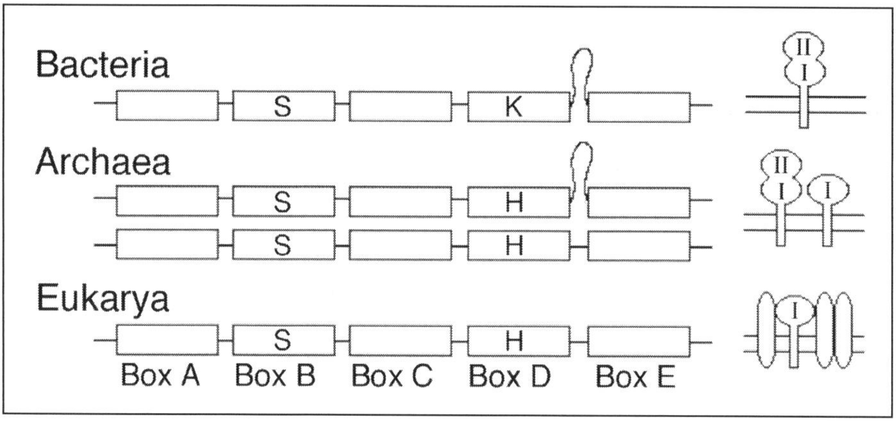

Figure 2. Comparison of type I signal peptidase across evolution. Schematic representation of signal peptidase from Bacteria, Archaea and Eukarya. On the left, homology boxes A-E are shown, as are the conserved serine and lysine residues of the catalytic dyad of the bacterial enzyme, or their replacements in the archaeal and eukaryal enzymes. On the right, the structure (depicting domains I and II) and oligomeric status of the enzyme in each phylogenetic group is schematically shown.

counterpart. Analysis of available archaeal signal peptidase sequences also reveals that a small number of the archaeal enzymes encode for a region termed domain II.[15] Domain II corresponds to a stretch found between sequence homology boxes D and E, and that, in the *E. coli* enzyme, folds into a large structure positioned on top of the catalytic core formed by boxes B-E.[93] The function of domain II and the reason for it being present in only some archaeal signal peptidases remains unclear. Still, the fact that some archaeal signal peptidases (e.g., *Thermoplasma acidophilum*) contain a domain II region whereas others do not could reflect an evolutionary scenario in which primitive archaeal type I signal peptidases originally contained domain II, however, during subsequent diversification, this region may have been lost in many strains.

At the protein level, only limited number study on archaeal signal peptidases have been conducted. It one study, type I signal peptidase from the methanoarchaea *Methanococcus voltae* was cloned and then expressed and characterized in a bacterial host.[16] Following heat inactivation to eliminate background activity of the native bacterial enzyme, the heterologously-expressed archaeal signal peptidase was shown to effectively cleave a truncated version of the *M. voltae* surface-layer glycoprotein, a natural substrate of the enzyme in this strain. Signal peptidase activity in *H. volcanii* membranes has also been addressed in vitro (Fine A, Irihimovitch I, Konrad Z, Eichler J, unpublished observations), where the ability of the enzyme to cleave the signal peptide of a reporter precursor was shown to be unaffected by standard protease inhibitors, as previously reported for eukaryal and bacterial signal peptidases. The activity of the archaeal enzyme was also shown to be insensitive to 5*S*-penem, an anti-bacterial signal peptidase-specific reagent.[93]

The signal peptidase responsible for the removal of signal peptides from methanoarchaeal preflagellins, the precursor form of proteins comprising the flagella, has also been addressed.[97] Unlike secretory preproteins, archaeal preflagellins bear signal peptides similar to those found on the precursor forms of bacterial type IV pilins, cell surface-associated structures involved in a variety of cellular processes.[98] It was subsequently shown that prepilin type IV signal peptidases could also cleave signal peptides from sugar binding proteins, and possibly other proteins, in *S. solfataricus*.[19] The reason why such signal peptides would be employed by non-flagellar exported *S. solfataricus* preproteins is unclear.

The Driving Force of Archaeal Protein Translocation

For a complete portrayal of protein translocation in Archaea, a description of the driving force of the system is required. However, as is the case for many aspects of archaeal protein translocation, little is known of the energetic considerations of the process. In a cotranslational translocation model, GTP-dependent elongation of ribosome-associated nascent polypeptides could provide the driving force for translocation. Such nascent chain-bearing ribosomes would be delivered to the translocon via the SRP targeting pathway, in a manner similar to what takes in Eukarya. By contrast, the driving force of post-translational archaeal translocation is more difficult to predict. Given the unlikelihood that elevated levels of ATP would be sequestered at the cell surface, it is improbable that Archaea rely on chaperones to pull polypeptides out of the cytoplasm, as is in the case in post-translational translocation into the ER or mitchondria. Indeed, many archaeal species fail to encode Hsp70 proteins,[99] a family of molecular chaperones that act to coordinate ATP hydrolysis with post-translational translocation in various systems.[100-104] As noted above, genome-based searches have failed to detect an archaeal SecA homologue. Given the high degree of conservation amongst bacterial and chloroplast SecA sequences, the apparent absence of an archaeal SecA homologue would argue against a Bacteria-like post-translational translocation process in Archaea. The existence of an archaeal structural homologue of SecA, undetectable through current sequence-based searches, however, cannot be discounted at this time.

Conclusions

Thus far, preliminary steps have been made towards a detailed description of archaeal protein translocation at the genetic, biochemical, structural and cell biology levels. Such efforts will surely benefit from the upcoming release of additional archaeal genome sequences and by investigations into other aspects of archaeal biology. Indeed, as new and improved molecular tools for working with a wide range of archaeal strains become available, it should become possible to reconstitute archaeal protein translocation in vitro. The clearer picture of archaeal protein translocation that will come from such studies will not only advance our understanding of the translocation process across evolution, but will also help decipher the molecular strategies adopted by extremophilic organisms in overcoming the challenges of their environments. Finally, the availability of a well-characterized archaeal protein export system will hasten realization of the enormous commercial potential associated with the large-scale production of industrially-useful extremophilic archaeal proteins.

Acknowledgements

Support comes from the Israel Science Foundation (grant 433/03).

References

1. Woese CR, Kandler O, Wheelis ML. Towards a natural system of organisms: Proposal for the domains Archaea, Bacteria and Eucarya. Proc Natl Acad Sci USA 1990; 87:4576-4579.
2. Graham DE, Overbeek R, Olsen GJ et al. An archaeal genomic signature. Proc Natl Acad Sci USA 2000; 97:3304-3308.
3. DeLong EF. Everything in moderation: archaea as 'non-extremophiles'. Curr Opin Genet Dev 1998; 8:649-654.
4. Rothschild LJ, Manicinelli RL. Life in extreme environments. Nature 2001; 409:1092-1101.
5. Rapoport TA, Jungnickel B, Kutay U. Protein transport across the eukaryotic endoplasmic reticulum and bacterial inner membrane. Annu Rev Biochem 1996; 65:271-303.
6. Johnson AE, van Waes MA. The translocon: a dynamic gateway at the ER membrane. Annu Rev Cell Dev Biol 1999; 15:799-842.

7. Manting EK, Driessen AJM. Escherichia coli translocase: the unravelling of a molecular machine. Mol. Microbiol. 2000; 37:226-238.
8. Ring G, Eichler J. Extreme secretion: Protein translocation across the archaeal plasma membrane. J Bioenerg Biomembr 2004; 36:35-45.
9. Pohlschroder M, Dilks K, Hand N et al. Translocation of proteins across archaeal cytoplasmic membranes. FEMS Microbiol Rev 2004; 28:3-24.
10. Eichler J. Archaeal protein translocation crossing membranes in the third domain of life. Eur J Biochem 2000; 267:3402-3412.
11. Jorgensen S, Vorgias CE, Antranikian G. Cloning, sequencing, characterization, and expression of an extracellular alpha-amylase from the hyperthermophilic archaeon Pyrococcus furiosus in Escherichia coli and Bacillus subtilis. J Biol Chem 1997; 272:16335-16342.
12. Horlacher R, Xavier KB, Santos H et al. Archaeal binding protein-dependent ABC transporter: molecular and biochemical analysis of the trehalose/maltose transport system of the hyperthermophilic archaeon Thermococcus litoralis. J Bacteriol 1998; 180:680-689.
13. Duffner F, Bertoldo C, Andersen JT et al. A new thermoactive pullulanase from Desulfurococcus mucosus: cloning, sequencing, purification, and characterization of the recombinant enzyme after expression in Bacillus subtilis. J Bacteriol 2000; 182:6331-6338.
14. Smith JD, Robinson AS. Overexpression of an archaeal protein in yeast: secretion bottleneck at the ER. Biotechnol Bioeng 2002; 79:713-723.
15. Eichler J. Archaeal signal peptidases from the genus Thermoplasma: structural and mechanistic hybrids of the bacterial and eukaryal enzymes. J Mol Evol 2002; 54:411-415.
16. Ng SY, Jarrell KF. Cloning and characterization of archaeal type I signal peptidase from Methanococcus voltae. J Bacteriol 2003; 185:5936-5942.
17. Saleh MT, Fillon M, Brennan PJ et al. Identification of putative exported/secreted proteins in prokaryotic proteomes. Gene 2001; 269:195-204.
18. Nielsen H, Engelbrecht J, Brunak S et al. Identification of prokaryotic and eukaryotic signal peptides and prediction of their cleavage sites. Protein Eng 1997; 10: 1-6.
19. Albers SV, Driessen AJM. Signal peptides of secreted proteins of the archaeon Sulfolobus solfataricus: a genomic survey. Arch Microbiol 2002; 177:209-216.
20. Bardy SL, Eichler J, Jarrell KF. Archaeal signal peptides—a comparative survey at the genome level. Protein Sci 2003; 12:1833-1843.
21. Faguy DM, Jarrell KF, Kuzio J et al. Molecular analysis of archaeal flagellins: similarity to the type IV pilin-transport superfamily widespread in bacteria. Can J Microbiol 1994; 40:67-71.
22. Rose RW, Bruser T, Kissinger JC et al. Adaptation of protein secretion to extremely high-salt conditions by extensive use of the twin-arginine translocation pathway. Mol Microbiol 2002; 45:943-950.
23. Dilks K, Rose RW, Hartmann E et al. Prokaryotic utilization of the twin-arginine translocation pathway: a genomic survey. J Bacteriol 2003; 185:1478-1483.
24. Bolhuis A. Protein transport in the halophilic archaeon Halobacterium sp. NRC-1: a major role for the twin-arginine translocation pathway? Microbiology 2002; 148:3335-3346.
25. Gropp R, Gropp F, Betlach MC. Association of the halobacterial 7S RNA to the polysome correlates with expression of the membrane protein bacterioopsin. Proc Natl Acad Sci USA 1992; 89:1204-1208.
26. Dale H, Angevine CM, Krebs MP. Ordered membrane insertion of an archaeal opsin in vivo. Proc Natl Acad Sci USA 2000; 97:7847-7852.
27. Dale H, Krebs MP. Membrane insertion kinetics of a protein domain in vivo. The bacterioopsin n terminus inserts co-translationally. J Biol Chem 1999; 274:22693-22698.
28. Ortenberg R, Mevarech M. Evidence for post-translational membrane insertion of the integral membrane protein bacterioopsin expressed in the heterologous halophilic archaeon Haloferax volcanii. J Biol Chem 2000; 275:22839-22846.
29. Irihimovitch V, Eichler J. Post-translational secretion of fusion proteins in the halophilic archaeon Haloferax volcanii. J Biol Chem 2003; 278:12881-12887.

30. Herskovits AA, Bibi E. Association of Escherichia coli ribosomes with the inner membrane requires the signal recognition particle receptor but is independent of the signal recognition particle. Proc Natl Acad Sci USA 2000; 97:4621-4626.
31. Keenan RJ, Freymann DM, Stroud RM et al. The signal recognition particle. Annu Rev Biochem 2001; 70:755-775.
32. Zwieb C, Eichler J. Getting on target: The archaeal signal recognition particle. Archaea 2001; 1:27-34.
33. Bhuiyan SH, Gowda K, Hotokezaka H et al. Assembly of archaeal signal recognition particle from recombinant components. Nucleic Acids Res. 2000; 28:1365-1373.
34. Diener JL, Wilson C. Role of SRP19 in assembly of the Archaeoglobus fulgidus signal recognition particle. Biochemistry 2000; 39:12862-12874.
35. Maeshima H, Okuno E, Aimi T et al. An archaeal protein homologous to mammalian SRP54 and bacterial Ffh recognizes a highly conserved region of SRP RNA. FEBS Lett 2001; 507:336-340.
36. Hainzl T, Huang S, Sauer-Eriksson AE. Structure of the SRP19 RNA complex and implications for signal recognition particle assembly. Nature 2002; 417:767-771.
37. Oubridge C, Kuglstatter A, Jovine L et al. Crystal structure of SRP19 in complex with the S domain of SRP RNA and its implication for the assembly of the signal recognition particle. Mol Cell 2002; 9:1251-1261.
38. Tozik I, Huang Q, Zweib C et al. Reconstitution of the signal recognition particle of the halophilic archaeaon Haloferax volcanii. Nucleic Acids Res. 2002; 30:4166-4175.
39. Moll R, Schmidtke S, Schäfer G. Domain structure, GTP-hydrolyzing activity and 7S RNA binding of Acidianus ambivalens Ffh-homologous protein suggest an SRP-like complex in archaea. Eur J Biochem 1999; 259:441-448.
40. Moll RG. Protein-protein, protein-RNA and protein-lipid interactions of signal-recognition particle components in the hyperthermoacidophilic archaeon Acidianus ambivalens. Biochem J 2003; 374:247-254.
41. Luirink, J, ten Hagen-Jongman CM, van der Weijden CC et al. An alternative protein targeting pathway in Escherichia coli: studies on the role of FtsY. EMBO J 1994; 13:2289-2296.
42. Lichi T, Ring G, Eichler J. Membrane binding of SRP pathway components in the halophilic archaea Haloferax volcanii. Eur J Biochem, 2004; 271:1382-1390.
43. Zelazny A, Seluanov A, Cooper A et al. The NG domain of the prokaryotic signal recognition particle receptor, FtsY, is fully functional when fused to an unrelated integral membrane polypeptide. Proc Natl Acad Sci USA 1997; 94:6025- 6029.
44. de Leeuw E, Poland D, Mol O et al. Membrane association of FtsY, the E. coli SRP receptor. FEBS Lett 1997; 416:225-229.
45. Powers T, Walter P. Co-translational protein targeting catalyzed by the Escherichia coli signal recognition particle and its receptor. EMBO J 1997; 16:4880-4886.
46. Connolly T, Gilmore R. GTP hydrolysis by complexes of the signal recognition particle and the signal recognition particle receptor. J Cell Biol 1993; 123:799-807.
47. Miller JD, Wilhelm H, Gierasch L et al. GTP binding and hydrolysis by the signal recognition particle during initiation of protein translocation. Nature 1993; 366:351-354.
48. Görlich D, Prehn S, Hartmann E et al. A mammalian homolog of SEC61p and SECYp is associated with ribosomes and nascent polypeptides during translocation. Cell 1994; 71:489-503.
49. Kalies KU, Görlich D, Rapoport TA et al. Binding of ribosomes to the rough endoplasmic reticulum mediated by the Sec61p-complex. J Cell Biol 1994; 126:925-934.
50. Prinz A, Behrens C, Rapoport TA et al. Evolutionarily conserved binding of ribosomes to the translocation channel via the large ribosomal RNA. EMBO J 2000; 19:1900-1906.
51. Zito CR, Oliver D. Two-stage binding of SecA to the bacterial translocon regulates ribosome-translocon interaction. J Biol Chem 2003; 278:40640-40646.
52. Ring G, Eichler J. Membrane binding of ribosomes occurs at SecYE-based sites in the Archaea Haloferax volcanii. J Mol Biol 2004; 336:997-1010.
53. Borgese N, Mok W, Kreibich G et al. Ribosomal-membrane interaction: in vitro binding of ribosomes to microsomal membranes. J Mol Biol 1974; 88:559-580.
54. Christian JHB, Waltho JA. Solute concentrations within cells of halophilic and non-halophilic bacteria. Biochem Biophys Acta 1962; 65:506-508.

55. Ginzburg M, Sachs L, Ginzburg BZ. Ion metabolism in a Halobacterium. I. Influence of age of culture on intracellular concentrations. J Gen Physiol 1970; 55:187-207.
56. Gorlich D, Rapoport TA. Protein translocation into proteoliposomes reconstituted from purified components of the endoplasmic reticulum membrane. Cell 1993; 75:615-630.
57. Brundage L, Hendrick JP, Schiebel E et al. The purified E. coli integral membrane protein SecY/E is sufficient for reconstitution of SecA-dependent precursor protein translocation. Cell 1990; 62:649-657.
58. Auer J, Spicker G, Bock A. Presence of a gene in the archaebacterium Methanococcus vannielii homologous to secY of eubacteria. Biochimie 1991; 73:683-688.
59. Kath T, Schäfer G. A secY homologous gene in the crenarchaeon Sulfolobus acidocaldarius. Biochim Biophys Acta 1995; 1264:155-158.
60. Cao TB, Saier MH Jr. The general protein secretory pathway: phylogenetic analyses leading to evolutionary conclusions. Biochim. Biophys. Acta 2003; 1609:115-125.
61. Irihimovitch V, Ring G, Elkayam T et al. Isolation of fusion proteins containing SecY and SecE components of the protein translocation complex from the halophilic archaeon Haloferax volcanii. Extremophiles 2003; 7:71-77.
62. Van den Berg B, Clemons WM Jr, Collinson I et al. X-ray structure of a protein-conducting channel. Nature 2004; 427:36-44.
63. Rensing SA, Maier U-G. The SecY protein family: comparative analysis and phylogenetic relationships. Mol Phylogen Evol 1994; 3:187-191.
64. Hartmann E, Sommer T, Prehn S et al. Evolutionary conservation of components of the protein translocation complex. Nature 1994; 367:654-657.
65. Brundage L, Fimmel CJ, Mizushima S et al. SecY, SecE, and band 1 form the membrane-embedded domain of Escherichia coli preprotein translocase. J Biol Chem 1992; 267:4166-4170.
66. Nishiyama K, Mizushima S, Tokuda H. A novel membrane protein involved in protein translocation across the cytoplasmic membrane of Escherichia coli. EMBO J 1993; 12:3409-3415.
67. Douville K, Leonard M, Brundage L et al. Band 1 subunit of Escherichia coli preportein translocase and integral membrane export factor P12 are the same protein. J Biol Chem 1994; 269:18705-18707.
68. Matlack KE, Mothes W, Rapoport TA. Protein translocation: tunnel vision. Cell 1998; 92:381-390.
69. Hanada M, Nishiyama KI, Mizushima S et al. Reconstitution of an efficient protein translocation machinery comprising SecA and the three membrane proteins, SecY, SecE, and SecG (p12). J Biol Chem 1994; 269:23625-23631.
70. Nishiyama K, Hanada M, Tokuda H. Disruption of the gene encoding p12 (SecG) reveals the direct involvement and important function of SecG in the protein translocation of Escherichia coli at low temperature. EMBO J 1994; 13:3272-3277.
71. Duong F, Wickner W. Distinct catalytic roles of the SecYE, SecG and SecDFyajC subunits of preprotein translocase holoenzyme. EMBO J 1997; 16:2756-2768.
72. Kalies KU, Rapoport TA, Hartmann E. The beta subunit of the Sec61 complex facilitates cotranslational protein transport and interacts with the signal peptidase during translocation. J Cell Biol 1998 ;141:887-894.
73. Kinch LN, Saier Jr MH, Grishin NV. Sec61beta-a component of the archaeal protein secretory system. Trends Biochem Sci 2002; 27:170-171.
74. Kates M. Membrane lipids of archaea. In: Kates M, Kushner DJ, Matheson AT, eds. The Biochemisty of Archaea (archaebacteria) NY: Elsevier, 1993:261-296.
75. Tseng TT, Gratwick KS, Kollman J. The RND permease superfamily: an ancient, ubiquitous and diverse family that includes human disease and development proteins. J Mol Microbiol Biotechnol 1999; 1:107-125.
76. Eichler J. Evolution of the prokaryotic protein translocation complex: a comparison of archaeal and bacterial versions of SecDF. Mol Phylogenet Evol 2003; 27:504-509.
77. Economou A, Pogliano JA, Beckwith J et al. SecA membrane cycling at SecYEG is driven by distinct ATP binding and hydrolysis events and is regulated by SecD and SecF. Cell 1995; 83:1171-1181.
78. Duong F, Wickner W. The SecDFyajC domain of preprotein translocase controls preprotein movement by regulating SecA membrane cycling. EMBO J 1997; 16:4781-4879.

79. Scotti PA, Urbanus ML, Brunner J et al. YidC, the Escherichia coli homologue of mitochondrial Oxa1p, is a component of the Sec translocase. EMBO J 2000; 19:542-549.
80. Moore M, Harrison MS, Peterson EC et al. Chloroplast Oxa1p homolog albino3 is required for post-translational integration of the light harvesting chlorophyll-binding protein into thylakoid membranes. J Biol Chem 2000; 275:1529-1532.
81. Hell K, Neupert W, Stuart RA. Oxa1p, an essential component of the N-tail protein export machinery in mitochondria. EMBO J 2001; 20:1281-1288.
82. Samuelson JC, Chen M, Jiang, F et al. YidC mediates membrane protein insertion in bacteria. Nature 2000; 406:637-641.
83. Chen M, Samuelson JC, Jiang F et al. Direct interaction of YidC with the Sec-independent Pf3 coat protein during its membrane protein insertion. J Biol Chem 2002; 277:7670-7675.
84. Luirink J, Samuelsson T, de Gier JW. YidC/Oxa1p/Alb3: evolutionarily conserved mediators of membrane protein assembly. FEBS Lett 20001; 501:1-5.
85. Yen MR, Tseng YH, Nguyen EH et al. Sequence and phylogenetic analyses of the twin-arginine targeting (Tat) protein export system. Arch Microbiol 2002; 177:441-450.
86. Chung YJ, Krueger C, Metzgar D et al. Size comparisons among integral membrane transport protein homologues in bacteria, Archaea, and Eucarya. J Bacteriol 2001; 183:1012-1021.
87. Yen MR, Harley KT, Tseng YH et al. Phylogenetic and structural analyses of the oxa1 family of protein translocases. FEMS Microbiol Lett 2001; 204:223-231.
88. Hartmann E, Gorlich D, Kostka S et al. A tetrameric complex of membrane proteins in the endoplasmic reticulum. Eur J Biochem 1993; 214:375-381.
89. Gorlich D, Hartmann E, Prehn S et al. A protein of the endoplasmic reticulum involved early in polypeptide translocation. Nature 1992; 357:47-52.
90. Berks BC, Sargent F, Palmer T. The Tat protein export pathway. Mol Microbiol 2000; 35:260-274.
91. Robinson C, Bolhuis A. 2001 Protein targeting by the twin-arginine translocation pathway. Nat Rev Mol Cell Biol 2001; 2:350-356.
92. Dalbey RE, Lively MO, Bron S et al. The chemistry and enzymology of the type I signal peptidases. Protein Sci 1997; 6:1129-1138.
93. Paetzel M, Dalbey RE, Strynadka NCJ. The structure and mechanism of bacterial type I signal peptidases. A novel antibiotic target. Pharmacol Ther 2000; 87:27-49.
94. Tjalsma H, Bolhuis A, van Roosmalen ML et al. Functional analysis of the secretory precursor processing machinery of Bacillus subtilis: identification of a eubacterial homolog of archaeal and eukaryotic signal peptidases. Genes Develop 1998; 12:2318-2331.
95. VanValkenburgh C, Chen X, Mullins C et al. The catalytic mechanism of endoplasmic reticulum signal peptidase appears to be distinct from most eubacterial signal peptidases. J Biol Chem 1999; 274:11519-11525.
96. YaDeau JT, Klein C, Blobel G. Yeast signal peptidase contains a glycoprotein and the Sec11 gene product. Proc. Natl. Acad. Sci. USA 1991; 88:517-521.
97. Correia JD, Jarrell KF. Posttranslational processing of Methanococcus voltae preflagellin by preflagellin peptidases of M. voltae and other methanogens. J Bacteriol 2000; 182:855-858.
98. Mattick JS. Type IV pili and twitching motility. Annu Rev Microbiol 2002; 56:289-314.
99. Macario AJ, Lange M, Ahring BK et al. Stress genes and proteins in the archaea. Microbiol Mol Biol Rev 1999; 63:923-967.
100. Wild J, Altman E, Yura T et al. DnaK and DnaJ heat shock proteins participate in protein export in Escherichia coli. Genes Dev 1992; 6:1165-1172.
101. Rial DV, Arakaki AK, Ceccarelli EA. Interaction of the targeting sequence of chloroplast precursors with Hsp70 molecular chaperones. Eur J Biochem 2000; 267:6239-6248.
102. Harano T, Nose S, Uezu R et al. Hsp70 regulates the interaction between the peroxisome targeting signal type 1 (PTS1)-receptor Pex5p and PTS1. Biochem J 2001; 357:157-165.
103. Ngosuwan J, Wang NM, Fung KL et al. Roles of cytosolic Hsp70 and Hsp40 molecular chaperones in post-translational translocation of presecretory proteins into the endoplasmic reticulum. J Biol Chem 2003; 278:7034-7042.
104. Young JC, Hoogenraad NJ, Hartl FU. Molecular chaperones Hsp90 and Hsp70 deliver preproteins to the mitochondrial import receptor Tom70. Cell 2003; 112:41-50.

Chapter 4

Structure of the SecYEG Protein Translocation Complex

Ian Collinson*

Abstract

Protein transport through and into biological membranes is a process of fundamental importance in all living organisms. In eukaryotes, protein translocation through the endoplasmic reticulum is carried out by a membrane protein complex called Sec61, usually while associated with ribosomes. In Bacteria and Archaea, protein translocation through and into the cytosolic membrane is conducted by the homologous SecY complex. In *Escherichia coli*, SecYEG consists of three polypeptides associating with either ribosomes or the partner ATPase SecA, which drive the translocation reaction. The structure of the SecY complex has been determined in a closed conformation. SecY encapsulates the central protein channel formed by the two halves of the subunit, closed by a short plug domain and a ring of hydrophobic residues. In combination with previous results, the atomic structure has led to models of how the complex might move during the reaction cycle, but events and conformational changes associated with the engagement of substrate and the partner protein are not understood. This chapter will review the recent structural results relating to the protein-conducting channel. The implications of these findings will be described in the context of the mechanism through which proteins pass across and into the membrane. The nature of the interaction with substrate and translocation partners will be discussed together with the possible movements that occur during the reaction cycle.

Proteins destined for secretion, membrane integration or organellar import contain signal sequences that direct them to the membrane. Once there, transport machines receive and translocate the substrate protein appropriately across or into the membrane. These essential reactions are controlled by an array of assemblies, which ensure that the proteins find the correct compartment. The Sec complex is the only protein channel conserved throughout biology.[1,2] Depending on the organism and type of substrate, the Sec complex cooperates with various partners to pull or push the substrate polypeptide by post- and cotranslational mechanisms.[2]

In Bacteria and Archaea, the SecY complex (SecYEG/SecYEβ) is composed of three membrane proteins and conducts proteins through or into the cytosolic membrane.[3] The *E. coli* SecY, SecE and SecG subunits each have 10, 3 and 2 trans-membrane domains,

*Ian Collinson—Department of Biochemistry, School of Medical Sciences, University of Bristol, U.K. Email: Ian.Collinson@bristol.ac.uk

Protein Movement Across Membranes, edited by Jerry Eichler. ©2005 Eurekah.com and Springer Science+Business Media.

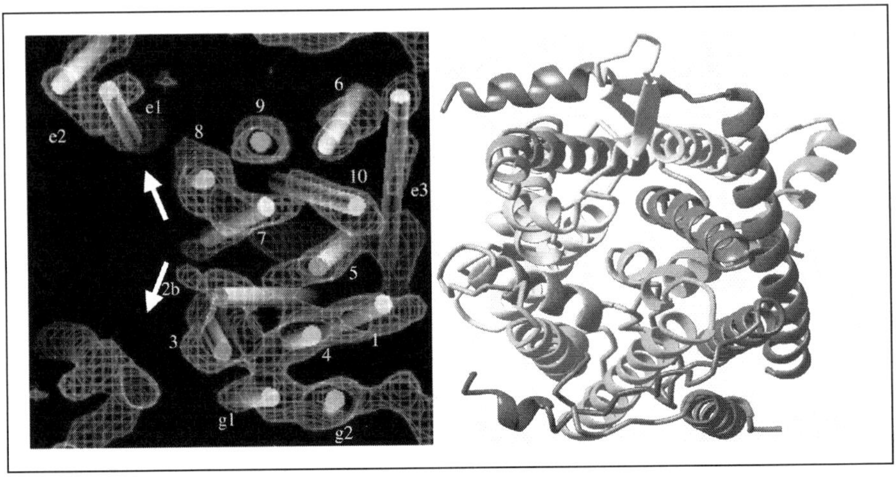

Figure 1. Comparison of the membrane-bound *E. coli* SecYEG and *M. jannaschii* SecYEβ structures. On the left side, the *E. coli* structure is shown from the cytoplasmic views with the helices numbered; SecY (1-10), SecE (e1-e3) and SecG (g1 and g2). The same view of the *M. jannaschii* complex is shown on the right. The latter complex lacks e1, e2 and g1. The white arrows indicate the possible movements of TM1-5 with respect to TM6-10 during the protein translocation reaction.

respectively. As in eukaryotes, the prokaryotic Sec complex can associate with substrates during their synthesis from the ribosome, which may then thread the polypeptide through the channel during protein synthesis.[4] Bacteria possess an additional pathway, whereby the newly synthesized substrate protein is maintained in an unfolded conformation, engaged by the ATPase SecA and delivered to the translocon.[5] A series of reactions follows, during which ATP is used to drive the passage of protein through the SecY complex.[6] The mechanism of this driving reaction remains the subject of investigation (see below). In contrast, the pathways followed by the translocating polypeptide chain and conformational changes that might be adopted by the channel have recently been clarified. Hence, concerning the SecY complex itself, there is now a wealth of available genetic, biochemical and structural data,[7,8] such that our understanding of the transport process through SecYEG is more advanced than of any other system.

The first pictures of the translocon were derived from electron micrographs of individual molecules; these were averaged to visualize ring-like structures of both the mammalian[9] and bacterial[10] complexes, each reminiscent of the other. Further studies showed the same pore complexes bound and aligned to the polypeptide exit site of the ribosome.[11-13] The resolution, however, was insufficient to resolve the individual Sec61 complexes of the assembly. Later crystallographic work resolved the complex in much more detail. The medium resolution structure of dimeric *E. coli* SecYEG was determined by electron cryo-microscopy,[7,14] while a detergent-solubilized monomer of the related SecYEβ from *M. jannaschii* was solved at atomic resolution by X-ray crystallography[8] (Fig. 1). The former structure was resolved from 2D-crystals, which could be imaged directly, and from the best flat and tilted images, the three-dimensional structure could be resolved by computer processing.[7] Importantly, the structure determined was in its native state, that is, bound to the membrane, meaning that the structure is likely to be in its active state. In practice, 2D-crystals are grown by the addition of detergent-solved lipids to the purified complex. Sometimes, detergent removal by dialysis results in the efficient incorporation of the complex into the

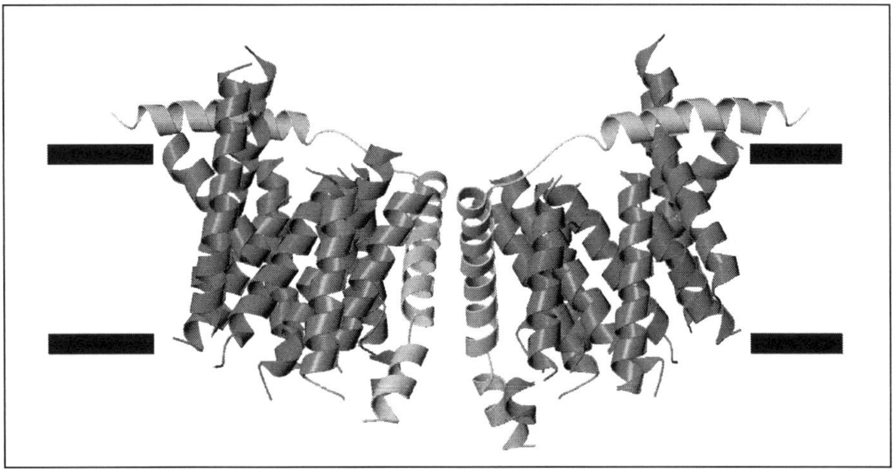

Figure 2. Model of the membrane-bound *E.coli* SecYEG dimer. The dimer is viewed from the side with its cytoplasmic surface uppermost shown. In dark grey is SecY, and in light grey are SecE and SecG.

bilayer such that under appropriate conditions, crystalline patches form in the plane of the membrane.[14] These are the same procedures used for the reconstitution of the Sec complex prior to in vitro translocation activity measurements, except that the protein : lipid ratio is varied accordingly.[14] The structure obtained in this manner revealed a dimer of SecYEG complexes bound in a 'back-to-back' fashion (Figs. 1 and 2).

As is typical for 2D-crystallographic projects, the resolution obtained is lower than is obtainable from classical X-ray diffraction analysis of 3D-crystals. As 2D-crystals are imaged using microscopy, both amplitude and phase structure factors are measured directly, alleviating the need to solve the phase problem. This means, unlike in the case of X-ray crystallography, that low and medium resolution structures can be readily solved. The attainable resolution is restricted by the crystals themselves and the nature of the data collection, which limits further the resolution perpendicular to the membrane. In spite of these drawbacks, the detail in the map was good enough to identify all of the trans-membrane α-helices, but not to assign them or to locate the positions of individual amino acids and their side chains. At the dimer interface there was a cavity open to the cytoplasmic face and closed by two highly tilted helices from each monomer, identified correctly as the essential trans-membrane domain 3 of SecE.[7] The same helices have been shown to efficiently cross-link to one another,[15] consistent with their close proximity at the dimer interface. There is also an area of low density surrounded by bundles of tilted trans-membrane helices in the middle of each monomer, which turned out to be the closed protein channel. Two peripheral outer lying helices were later identified as the nonessential N-terminal trans-membrane domains 1 and 2 of SecE.

The second translocon structure was determined by X-ray crystallography of a related complex from *Methanococcus jannaschii*.[8] The resolution obtained was 3.2 Å, enough to build an atomic model of the complex (Fig. 1). The structure determined was of the monomeric form from 3D-crystals grown in detergent solution. This is a radically different approach than that employed for 2D-crystallography, described above. Here, high concentrations of detergent are present in the crystallization mix and the protein is likely stripped of all lipids. Combined, these effects may have perturbed the structure of the complex, especially given that the

E. coli complex is known to dissociate from its active dimeric state to the monomeric form upon removal of lipids and addition of high concentrations of detergent.[16] As there has been little functional analysis of the archaeal complex either in vivo or in vitro, the SecYEβ preparation was not assayed for translocation activity. The availability of a structure of the *E. coli* homologue for comparison was, therefore, important. The two structures are indeed very similar, indicating that the high-resolution crystal structure is meaningful and reflects an active-like state. However, the archaeal structure is monomeric, in contrast to the low- and medium-resolution structures of the mammalian and bacterial Sec61/Y complexes.

Both the 2D and 3D structures show the protein conduction channel, which is held closed between two pseudo-symmetric domains of the large SecY subunit. The channel, formed in the centre of a monomeric complex, is blocked by a short plug domain and a ring of hydrophobic residues.[8] The channel lining is quite well-conserved and also forms a hotspot for a collection of *prl* mutants that have effects (usually positive) on the translocation reaction.[8] The structure of the SecY complex also revealed the location of the signal sequence binding site, identified by cross-linking to lie between trans-membrane domains TM2b and 7 of SecY.[17] The site is situated adjacent to the channel and between the two lobes formed by each half of the SecY subunit.[8] An analysis of the size of the protein pore in the active mammalian translocon revealed that it might open up to a diameter of up to 60 Å.[18] If true, this is an incredibly large opening requiring enormous conformational changes within the complex.

The mechanism of channel gating and the rearrangements required to bind and translocate protein are not clear. The channel might open by the movement of the N- and C- terminal domains of SecY like a 'crab's claw' about a hinge between the two halves (Fig. 1). SecY is clamped together by a peripheral subunit SecE, holding the structure closed. The two additional peripheral trans-membrane helices of SecE look like they might reinforce its embrace around the two potentially moving halves of SecY.[7,8] In addition, it appears that the displacement of the plug to a new position in the structure would be required for translocation to occur. These movements are probably different, depending on whether the complex transports substrate domains through or into the lipid bilayer. It should be noted, however, that although these models are based on the high-resolution structure of the complex,[8] they lack detail. Indeed, several of the implications remain speculative, particularly because the channel was crystallized in its closed state, in its (inactive) monomeric form and without its interacting partners. The proposed movement of the plug away from the central blocking position to a location close to the C-terminus of SecE nonetheless does have experimental support from the identification of an in vivo cross-link of the plug to this open position.[19]

The structures of ribosomes[20,21] and of SecA,[22-24] both reaction partners of the Sec complexes, have also been determined. In spite of these advances, not much is known about the interaction between them. The low resolution structures of ribosome-Sec61 complexes lack the detail required to localize and examine the sites at the interface.[12] Archaea do not possess the SecA partner and thus may rely more on the cotranslational mode of protein translocation for their Sec-dependent export requirements. The large cytosolic loops of *M. jannaschii* SecYEβ presumably involved in interactions with ribosomes are visible, but how they do so is unclear.

In Bacteria, the nature of the interaction of SecYEG with the partner SecA during post-translational translocation is also not fully understood. The recent determination of the atomic structure of monomeric SecYEβ has allowed for the building of an *E. coli* homology model (Bostina, Mohsin, Kühlbrandt, Collinson, unpublished results). In this structure, the predicted positions of the large two cytoplasmic loops and C-terminus of SecY are ideally poised for an interaction with an approaching SecA.

Various models have been proposed to describe the mechanism of SecA-driven protein translocation.[6,22] A multi-step process involving several rounds of ATP hydrolysis by SecYEG and membrane-associated SecA thrusts the polypeptide chain through the protein channel.[25] For each round of the reaction, it is thought that about 5 kDa of protein is pushed into the complex and through the membrane.[26] There have been suggestions that SecA at least partially inserts itself (and substrate protein) through the SecYEG complex.[27,28] Based on these and other findings, models have been proposed that translocation occurs by the threading of the chain across the membrane, in a manner reminiscent of a sewing machine.[6]

In its current state, the nature of the SecYEG structure, which lacks large internal cavities, seems to suggest that it would be unlikely for SecA or a 30 kDa domain of the protein to insert into the SecY complex or to traverse the entire membrane in this way. Like other channels and transporters, the Sec complex has effected a hydrophilic constriction of the membrane in order to minimize the considerable movements that must be required in a reaction like this. More likely, the conformational changes experienced by SecA will be subtler. In reality, we currently lack detail concerning the nature of the interactions and conformational changes that occur between these components during the early or later stages of the post-translational reaction in Bacteria. Reports of the enzymology and oligomeric state of the SecYEG-bound SecA have failed to produce a consensus on the basic mechanism of action; for example, the stoichiometry of the associated partners has not been clearly defined.[29,30] This is partly because the interaction and the activity of the pair has only been addressed using reconstituted membranes and cannot easily be studied in solution. Recent work even suggests that the stoichiometry of the associated translocation pair might vary.[31,32] The latter study has identified two forms of a large translocation assembly: one, in the presence of the nonhydrolyzable ATP analogue, AMPPNP—$(SecYEG)_2SecA_2$; and another in the absence of added nucleotides—$(SecYEG)_2SecA_1$.[32] Whether or not this variability is important and varies during the reaction cycle remains to be seen.

The oligomeric state of the active SecY complex also lacks agreement; monomers,[33] dimers[16,32,34] and tetramers[34] have all been implicated in the translocation reaction. The apparent fact that an oligomeric form of the complex is found in membranes and is active[7,9,31,32,35] and yet the monomeric form of the SecY complex constitutes the protein channel[8] is a paradox that remains unsolved. It may be that the environment of the membrane promotes the oligomerization of the complex, as is the case for the *E. coli* complex.[7,9] A dimeric and back-to-back arrangement is consistent with the lateral release mechanism for integrating trans-membrane domains, as the two gateways face away from one another and toward the lipid bilayer. The experimentally-determined *E. coli* structure has been used as a guide to fit the high resolution *M. jannaschii* structure (Bostina, Mohsin, Kühlbrandt, Collinson, unpublished results), allowing for this arrangement of dimers to be reconstructed. A side view of the atomic model of this dimeric arrangement of membrane-bound *E. coli* SecYEG is shown in Figure 2. It should be noted that another protein factor, YidC, is absolutely essential for the integration of a handful of membrane proteins.[36] YidC seems to be loosely-associated with the translocon[37,38] and might, therefore, bind to the proposed sites of the lateral release of gateways to facilitate the partitioning of trans-membrane α-helices into the lipid phase of the membrane. The arrangement of dimers would be consistent with this idea.

Tetramers of the SecY complex have also been observed.[14,16,34] In addition, Driessen and coworkers observed a SecA-induced tetramerization of the complex.[34] Elsewhere, however, neither the monomer nor tetramer associate with SecA,[32] with tetramers only being visualized when the complex was present at unnaturally high concentrations.[14,16] In contrast, the ribosome-bound complex appears as though it is larger than a dimer.[12] An earlier study also reported that the Sec61 complex oligomerizes in membranes following the addition of ribosomes. Such oligomers were reported to be made of 3-4 copies of the complex.[9]

Given that the translocon channel seems to be active as an oligomer, it is unclear whether or why the presence of multiple active sites is necessary for activity. Stability is a possible explanation. The size of monomeric SecYEG is about 70 kDa, compared to 200 kDa for the SecA dimer, or ~2 MDa for the ribosome. Perhaps the SecYEG dimer or tetramer might be needed to provide a big enough platform for the association of these large molecules. Alternatively, the interaction between two SecYEG monomers at the dimeric interface may bring about conformational changes essential for SecA and substrate binding and for subsequent translocation reactions. There are some indications that this may indeed be the case (Bostina, Mohsin, Kühlbrandt & Collinson, unpublished results). Finally, there may be some unusual regulatory or allosteric features of translocation that could only be provided by more than one copy of the SecYEG heterotrimer. The apparent need and variability in the oligomeric associations and stoichiometry of SecYEG, SecA and ribosomes, although puzzling, might have important implications for the translocation reaction and may be a consequence of the need to suit different modes of translocation and substrate. The requirement for the initiation of translocation is likely to be different for post- and co-translocational modes and could explain why there is an apparent preference for dimers and tetramers, respectively. The contacts that occur at dimer or tetramer interfaces in the membrane would presumably be different and might modulate the structure of the Sec complex to allow for the translocation of different substrates driven by different partner complexes.

Future work will focus on these problems to further our understanding of the gating mechanism of the translocation reaction. The key will be to identify in detail the nature of interactions and dynamics that occur between the channel, partners and substrate protein during the engagement of the reaction. The consequences of this reaction, the conformational changes that occur and the energetics that drive the passage of protein appropriately through and into the membrane, are all presently poorly understood and require further insight.

Acknowledgements

Ian Collinson is supported by the EMBO Young Investigator Program and by a grant from the DFG Sonderforschungsbereich 628. Thanks to Dr. Mihnea Bostina for help with the figures.

References

1. Hartmann E, Sommer T, Prehn S et al. Evolutionary conservation of components of the protein translocation complex. Nature 1994; 367:654-657.
2. Matlack KE, Mothes W, Rapoport TA. Protein translocation: Tunnel vision. Cell 1998; 92:381-390.
3. Douville K, Price A, Eichler J et al. SecYEG and SecA are the stoichiometric components of preprotein translocase. J Biol Chem 1995; 270:20106-20111.
4. Gorlich D, Prehn S, Hartmann E et al. A mammalian homolog of SEC61p and SECYp is associated with ribosomes and nascent polypeptides during translocation. Cell 1992; 71:489-503.
5. Brundage L, Hendrick JP, Schiebel E et al. The purified E. coli integral membrane protein SecY/E is sufficient for reconstitution of SecA-dependent precursor protein translocation. Cell 1990; 62:649-657.
6. Duong F, Eichler J, Price A et al. Biogenesis of the gram-negative bacterial envelope. Cell 1997; 91:567-573.
7. Breyton C, Haase W, Rapoport TA et al. Three-dimensional structure of the bacterial protein-translocation complex SecYEG. Nature 2002; 418:662-665.
8. van den Berg L, Clemons WMJ, Collinson I et al. X-ray structure of a protein-conducting channel. Nature 2004; 427:36-44.
9. Hanein D, Matlack K, Jungnickel B et al. Oligomeric rings of the Sec61p complex induced by ligands required for protein translocation. Cell 1996; 87:721-732.

10. Meyer T, Menetret J, Breitling R et al. The bacterial SecY/E translocation complex forms channel-like structures similar to those of the eukaryotic Sec61p complex. J Mol Biol 1999; 285:1789-1800.
11. Beckmann R, Bubeck D, Grassucci R et al. Alignment of conduits for the nascent polypeptide chain in the ribosome-Sec61 complex. Science 1997; 278:2123-2126.
12. Beckmann R, Spahn C, Eswar N et al. Architecture of the protein-conducting channel associated with the translating 80S ribosome. Cell 2001; 107:361-372.
13. Menetret J, Neuhof A, Morgan D et al. The structure of ribosome-channel complexes engaged in protein translocation. Mol Cell 2000; 6:1219-1232.
14. Collinson I, Breyton C, Duong F et al. Projection structure and oligomeric properties of a bacterial core protein translocase. EMBO J 2001; 20:2462-2471.
15. Veenendaal A, van der Does C, Driessen A. Mapping the sites of interaction between SecY and SecE by cysteine scanning mutagenesis. J Biol Chem 2001; 276:32559-32566.
16. Bessonneau P, Besson V, Collinson I et al. The SecYEG preprotein translocation channel is a conformationally dynamic and dimeric structure. EMBO J 2002; 21:995-1003.
17. Plath K, Mothes W, Wilkinson B et al. Signal sequence recognition in posttranslational protein transport across the yeast ER membrane. Cell 1998; 94:795-807.
18. Hamman B, Chen J, Johnson E et al. The aqueous pore through the translocon has a diameter of 40-60 A during cotranslational protein translocation at the ER membrane. Cell 1997; 89:535-544.
19. Harris CR, Silhavy TJ. Mapping an interface of SecY (PrlA) and SecE (PrlG) by using synthetic phenotypes and in vivo cross-linking. J Bacteriol 1999; 181:3438-3444.
20. Ban N, Nissen P, Moore PB et al. The complete atomic structure of the large ribosomal subunit at 2.4 A resolution. Science 2000; 289:905-920.
21. Yusupov MM, Yusupova GZ, Baucom A et al. Crystal structure of the ribosome at 5.5 A resolution. Science 2001; 292:883-896.
22. Hunt JF, Weinkauf S, Henry L et al. Nucleotide control of interdomain interactions in the conformational reaction cycle of SecA. Science 2002; 297:2018-2026.
23. Osborne AR, Clemons Jr WM, Rapoport TA. A large conformational change of the translocation ATPase SecA. Proc Natl Acad Sci USA 2004; 101:10937-10942.
24. Sharma V, Arockiasamy A, Ronning DR et al. Crystal structure of Mycobacterium tuberculosis SecA, a preprotein translocating ATPase. Proc Natl Acad Sci USA 2003; 100:2243-2248.
25. Bassilana M, Wickner W. Purified Escherichia coli preprotein translocase catalyzes multiple cycles of precursor protein translocation. Biochemistry 1993; 32:2626-2630.
26. van der Wolk JP, de Wit JG, Driessen AJ. The catalytic cycle of the Escherichia coli SecA ATPase comprises two distinct preprotein translocation events. EMBO J 1997; 16:7297-7304.
27. Economou A, Wickner W. SecA promotes preprotein translocation by undergoing ATP-driven cycles of membrane insertion and deinsertion. Cell 1994; 78:835-843.
28. Eichler J, Wickner W. Both an N-terminal 65-kDa domain and a C-terminal 30-kDa domain of SecA cycle into the membrane at SecYEG during translocation. Proc Natl Acad Sci USA 1997; 94:5574-5581.
29. Driessen A. SecA, the peripheral subunit of the Escherichia coli precursor protein translocase, is functional as a dimer. Biochemistry 1993; 32:13190-13197.
30. Or E, Navon A, Rapoport TA. Dissociation of the dimeric SecA ATPase during protein translocation across the bacterial membrane. EMBO J 2002; 21:4470-4479.
31. Duong F. Binding, activation and dissociation of the dimeric SecA ATPase at the dimeric SecYEG translocase. EMBO J 2003; 22:4375-4384.
32. Tziatzios C, Schubert D, Lotz M et al. The bacterial protein-translocation complex: Dimeric SecYEG associates with both one or two molecules of SecA. J Mol Biol 2004; 340:513-524.
33. Yahr T, Wickner W. Evaluating the oligomeric state of SecYEG in preprotein translocase. EMBO J 2000; 19:4393-4401.
34. Manting EH, van der Does C, Remigy H et al. SecYEG assembles into a tetramer to form the active protein translocation channel. EMBO J 2000; 19:852-861.
35. Mori H, Tsukazaki T, Masui R et al. Fluorescence resonance energy transfer analysis of protein translocase. SecYE from Thermus thermopholus HB8 forms a constitutive oligomer in membranes. J Biol Chem 2003; 278:14257-14264.

36. Samuelson J, Chen M, Jiang F et al. YidC mediates membrane protein insertion in bacteria. Nature 2000; 406:637-641.
37. Duong F, Wickner W. Distinct catalytic roles of the SecYE, SecG and SecDFyajC subunits of preprotein translocase holoenzyme. EMBO J 1997; 16:2756-2768.
38. Scotti P, Urbanus M, Brunner J et al. YidC, the Escherichia coli homologue of mitochondrial Oxa1p, is a component of the Sec translocase. EMBO J 2000; 19:542-549.

CHAPTER 5

Membrane Protein Insertion in Bacteria from a Structural Perspective

Mark Paetzel and Ross E. Dalbey*

Abstract

Membrane proteins are inserted into the lipid bilayer in Bacteria by two pathways. The Sec machinery is responsible for the insertion of the majority of the membrane proteins after targeting by the SRP/FtsY components. However, there is also a class of membrane proteins that insert independent of the Sec machinery. These proteins require a novel protein called YidC. Recently, the structural details of the Sec machinery have come to light via X-ray crystallography. There are now structures of the membrane-embedded Sec protein-conducting channel, the SecA ATPase motor, and the targeting components. Structural information gives clues to how a polypeptide is translocated across the membrane and how the transmembrane segments of a membrane protein are released from the Sec complex. Additionally, the structures of the targeting components shed light on how substrates are selected for transport and delivered to the membrane.

Introduction

Membrane proteins are ubiquitous in nature and comprise around 30% of the total proteins within the cell. Membrane proteins play vital functions for the cell. They act as receptors where they are involved in transmitting information from the extracellular environment into the interior of the cell. Membrane proteins also function as transporters to move sugars, amino acids and other energy-rich molecules and ions into the cell. Other functions of membrane proteins include energy harvesting and energy transduction roles in photosynthesis and oxidative phosphorylation, as well as functions in lipid synthesis and catabolism. Given the wide variety of functions, there is a diversity of membrane protein structures. However, generally almost all integral membrane proteins in the inner membrane of Bacteria have helical transmembrane segments that range from 20 to 30 residues in length, with tryptophan and tyrosine residues being enriched near phospholipid headgroups and the connecting loops between helical transmembrane segments tend to be short.[1] In this review, we will bring the reader up to date on the latest developments in bacterial membrane protein biogenesis with a focus on structural aspects of the targeting and translocation components that facilitate insertion.

In the field of membrane protein biogenesis, there are at least four main problems: (1) How do membrane proteins with hydrophobic surfaces avoid aggregating in the cytoplasm?

*Corresponding Author: Ross E. Dalbey—Department of Chemistry, The Ohio State University, Columbus, Ohio 43210, U.S.A. Email: Dalbey@chemistry.ohio-state.edu

Protein Movement Across Membranes, edited by Jerry Eichler. ©2005 Eurekah.com and Springer Science+Business Media.

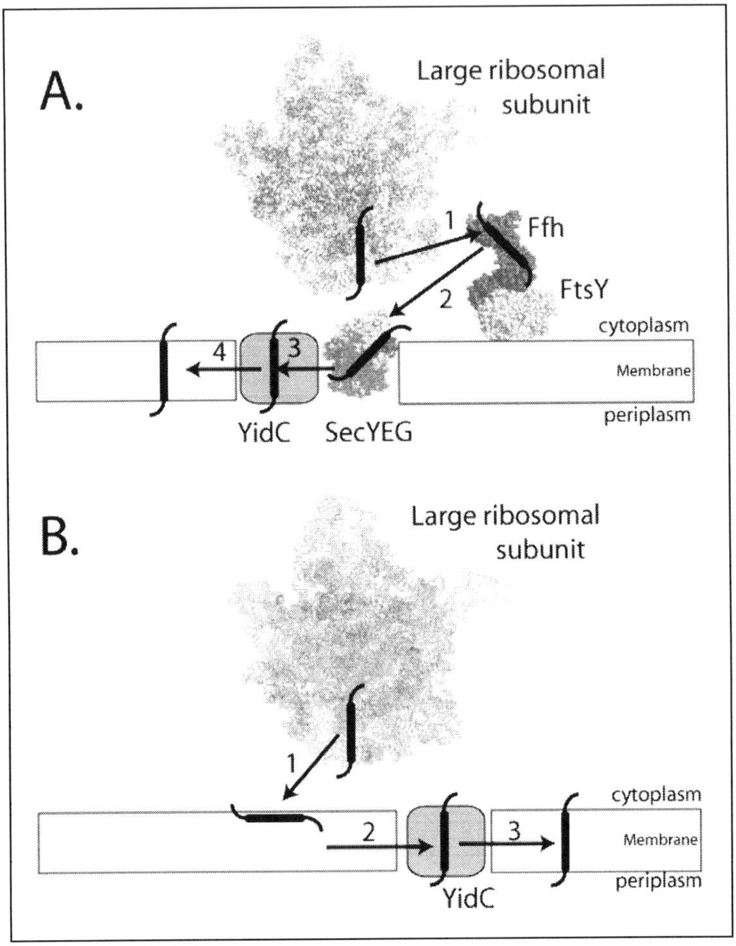

Figure 1. Schematic depiction of the two known membrane protein integration (assembly) pathways. A) The Sec-dependent pathway (the heterotrimer SecDFyajC and the ATPase SecA are not shown). B) The YidC pathway. The PDB coordinates used for the large ribosomal subunit from *Deinococcus radiodurans*[65] were 1NKW, the PDB coordinates used for SecYEβ from *Methanococcus janaschii*,[47] were 1RHZ. The PDB coordinates used for Ffh from *Sulfolobus solfataricus*[16] and FtsY from *Thermus aquaticus*[21] were 1QZW and 1RJ9, respectively. The program PyMol[66] was used to make this figure.

(2) How are hydrophilic domains translocated across the membrane? (3) How are hydrophobic domains integrated into the membrane? (4) What are the energetics of membrane protein insertion? Not surprising, there are proteins that catalyze the targeting of proteins to the membrane and the insertion into the lipid bilayer. In Bacteria, there are two pathways used for membrane protein insertion; the Sec pathway and YidC pathway. The majority of proteins use the Sec pathway for insertion (Fig. 1A). A subset of proteins insert by a Sec-independent pathway involving YidC (Fig. 1B).

The goal of understanding the molecular events involved in membrane protein assembly is not only of significant scientific interest in the membrane biogenesis area but is essential for the understanding of the disease states that result when these events go wrong.[2,3]

Insertion by the Sec Translocase-Mediated Pathway

Many membrane proteins inserted by the Sec pathway are targeted to the membrane by the evolutionarily-conserved Signal Recognition Particle (SRP) route. In this pathway, the cytosolic component SRP, comprised of Ffh and the 4.5S RNA,[4,5] binds to the membrane protein and targets the protein to the SRP receptor, FtsY. SRP binds to the hydrophobic region of the membrane protein as it emerges from the ribosomal tunnel (Fig. 1A). Then, the ribosome/mRNA/nascent membrane protein/Ffh complex is targeted to FtsY associated with the membrane.

Insertion of a protein into the membrane is initiated by a cleavable signal peptide or a noncleaved transmembrane segment. The transmembrane segments are integrated into the membrane and the hydrophilic domains are either translocated across the membrane or remain within the cytoplasm. The membrane protein uses the Sec translocase for insertion into the membrane and translocation of hydrophilic domains across the membrane (Fig. 1A). In *E. coli*, the Sec translocase is comprised of the SecYEG protein-conducting channel and the trimeric SecDFYajC complex (for review, see ref. 6). The protein YidC interacts with the hydrophobic regions of membrane proteins during the insertion of the protein into the membrane.[7] In some cases, the membrane-associated ATPase SecA is required for the translocation of large hydrophilic domains of membrane proteins.[8-10]

Targeting

The targeting components Ffh and FtsY are important for the insertion of membrane proteins as depletion of Ffh and FtsY within the cell has been shown to inhibit the insertion of a variety of membrane proteins. The SRP component Ffh in *E. coli* is homologous to the 54 kDa subunit of the eukaryotic SRP,[11] comprised of 6 polypeptides and a 7S RNA component.[5] Ffh exists in complex with a 4.5S RNA instead of the 7S RNA seen in the eukaryotic complex. SRP Ffh has been shown to bind to signal peptides of exported proteins and hydrophobic segments of membrane proteins.[10,12] For membrane proteins containing multiple hydrophobic regions, it may be sufficient for Ffh to bind to the first hydrophobic domain and target the protein to the membrane. Efficient membrane targeting of proteins which have hydrophobic surfaces is important as it prevents aggregation in the aqueous cytoplasm. The SRP receptor in Bacteria (FtsY) is simpler than the SRP receptor (SR) in eukaryotes, which contain two subunits, SRα and SRβ. The membrane-associated FtsY is homologous to the SRα subunit. Both FtsY and Ffh are essential bacterial proteins.[13,14] Ffh has been shown to form a complex with FtsY, in a GTP-dependent manner.[15] Following GTP hydrolysis, the Ffh and FtsY complex disassembles from the targeted nascent protein and the nascent chain can insert into the Sec machinery. Interestingly, it has been found that the GTPase activity of Ffh is stimulated by FtsY[15] while the GTPase activity of FtsY is stimulated by Ffh.

In order to provide insight into the protein targeting mechanism, it is very useful to obtain structural knowledge of the targeting components. Ffh contains three domains, i.e, the amino-terminal N domain, the GTPase G domain and the methionine-rich M domain (Fig. 2A).[16,17] The M domain is connected to the N and G domains by a flexible linker. The crystal structure of the M domain from *Thermus aquaticus* reveals a hydrophobic groove lined with methionine residues that has been proposed to bind to the signal peptide or the membrane anchor domain of the nascent polypeptide.[17] Interestingly, a crystal structure of the *E. coli* Ffh domain with domain IV of the 4.5S RNA suggests that the signal sequence recognition domain is comprised of both protein and RNA (SRP)(Table 1A).[18] A structure of the complete SRP54 (Ffh) in complex with helix 8 of the SRP RNA component revealed the overall juxtaposition of the M, G and N domains relative to each other.[16] Numerous

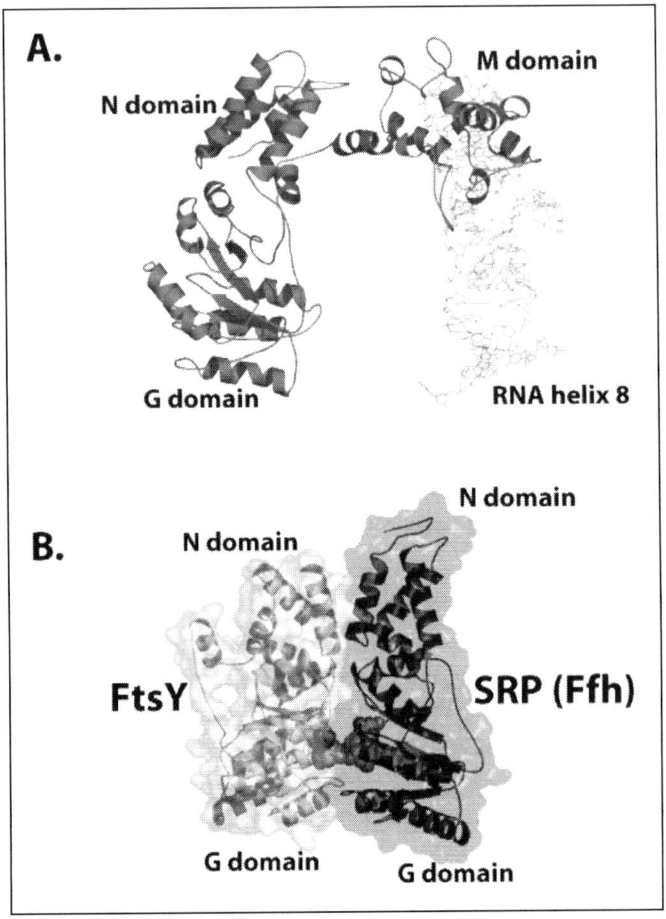

Figure 2. A) A ribbon diagram of the overall structure of the SRP core from the archaeon *Sulfolobus solfataricus*.[16] The structure reveals the interdomain communication between the N domain, the G domain, the M domain and helix 8 of SRP RNA. The RNA is shown in a stick diagram. The PDB coordinates 1QZW and the program PyMol[66] were used to make this figure. B) A ribbon diagram with transparent surface showing the heterodimeric complex of the signal recognition particle protein Ffh and its receptor FtsY from *Thermus aquaticus*.[21] Ffh is rendered in a darker shade and FtsY is shown in a lighter shade. The bound GTP analogue molecules are shown in van der Waal's spheres. The N-terminal domains (N domain) and the GTP binding domains (G domain) for each protein are labeled. The PDB coordinates 1RJ9 and the program PyMol[66] were used to make this figure.

structures are available for the NG domains of Ffh from archaeal homologs. These structures have been solved both in the presence and absence of GDP or non-hydrolyzable GTP analogs (see Table 1B). The N domain is comprised of a four-helix bundle, which is closely associated with the G domain (Ras-like GTPase) that has a core made up of a five-stranded β-sheet surrounded by α-helices. The G domain also contains an Insertion Box Domain (IBD) which is unique to the SRP GTPases. A similar structural arrangement is found in the N and G domains of *E. coli* FtsY (SRα), which has been solved to 2.2Å resolution.[19]

The structure of the catalytic core (N and G domains) formed by the Ffh/FtsY complex from *T. aquaticus* has been solved to 1.9 Å resolution in complex with the non-hydrolyzable

Table 1A. SRP protein/RNA complex structures

PDB ID	Source	Method	Description	R Value	Resolution [Å]	Reference
1DUL	*E. coli*	X-ray	Domain IV of 4.5 S RNA, M domain of Ffh	0.199	1.8	Batey et al 2000[18]
1HQ1	*E. coli*	X-ray	4.5S RNA, M domain of Ffh	0.151	1.5	Batey et al 2001 [67]
1QZW	*S. solfataricus*	X-ray	The complete SRP 54 (Ffh) with helix 8	0.340	4.1	Rosendal et al 2003[16]

GTP analog GMP-PCP.[20,21] The structures show that Ffh and FtsY form a quasi-two-fold symmetrical heterodimer having interaction surfaces both in the N domains and the G domains but with the majority of the protein-protein interactions occur between the G domains (Fig. 2B). Comparison with structures of the uncomplexed proteins shows there are major conformational changes that occur upon formation of the heterodimer. Binding of GTP versus GDP results in small structural adjustments in the free proteins.[22] The structures reveal that the 3' OH of the GTPs are essential for Ffh/FtsY association, activation and catalysis. The structures show that there is a shared composite active site containing the two GTPs at the interface, explaining why binding of Ffh to FtsY is GTP-dependent and why the complex disassembles after GTP hydrolysis. The structural rearrangement upon complex formation results in bringing catalytic residues in the IBD loop into the active site. The only interactions at the active site between the GTPases occur between the nucleotides. The GTP molecules are aligned head to tail such that the γ-phosphate of each GTP is hydrogen-bonded to the other GTP's ribose 3' OH group. Hydrolysis of the GTP releases the γ-phosphate. This essentially breaks the contact between the active sites and the GTP substrate and initiates the Ffh/FtsY dissociation. All the three-dimensional structural information for bacterial and archaeal SRP targeting components currently available is listed in Table 1A-C. The Signal Recognition Particle Database (SRPDB) (http://psyche.uthct.edu/dbs/SRPDB/SRPDB.html) provides up to date access to alignments of the SRP and SR sequences and phylogenic analysis of these proteins and RNAs.

The function of the SRP/FtsY domains become more clear upon structural analysis. Not only do the structures shed light on how the SRP Ffh M domain binds to the signal peptide, but they also deepen our understanding into why Ffh and FtsY act as each other's GTPase activating protein. The structures of the Ffh/FtsY (NG domain) complex reveal that Ffh and FtsY interact via the NG domains with the two GTPs forming a composite active site and explains why the targeting of ribosome nascent chain-bound Ffh to FtsY requires GTP (Fig. 5A). The transfer of the nascent membrane protein to the SecY complex cannot take place until Ffh bound to FtsY dissociates from the nascent chain. This only occurs after GTP has been hydrolyzed from Ffh and FtsY.

Translocation/Insertion

After targeting to the membrane, the hydrophobic signal anchor of the nascent membrane protein inserts into the SecYEG channel. The hydrophilic region of the membrane polypeptide is translocated through the Sec complex to the other side of the membrane and the membrane anchor region leaves the channel laterally. How the ribosome-bound membrane-targeted protein is transferred to the SecYEG channel is not known. One possibility is that there is a

Table 1B. SRP Ffh and FtsY protein structures

PDB ID	Source	Method	Description	R Value	Resolution [Å]	Reference
1FFH	*T. aquaticus*	X-ray	N and G domains of Ffh	0.186	2.0	Freymann et al 1997[68]
2FFH	*T. aquaticus*	X-ray	M domain of Ffh	0.257	3.2	Keenan et al 1998[17]
1NG1	*T. aquaticus*	X-ray	N and G domains of Ffh with GDP bound	0.189	2.0	Freymann et al 1999[69]
2NG1	*T. aquaticus*	X-ray	N and G domains of Ffh with GDP bound	0.200	2.0	Freymann et al 1999[69]
3NG1	*T. aquaticus*	X-ray	N and G domains of Ffh with no GDP bound	0.199	2.3	Freymann et al 1999[69]
1J8M	*A. ambivalens*	X-ray	G domain of Ffh	0.219	2.0	Montoya et al 2000[70]
1J8Y	*A. ambivalens*	X-ray	G domain of Ffh, T112A mutant	0.227	2.0	Montoya et al 2000[70]
1JPJ	*T. aquaticus*	X-ray	N and G domains of Ffh with the non-hydrolyzable GTP analog GMPPNP (N1 = crystal form 1)	0.201	2.3	Padmanabhan et al 2001[71]
1JPN	*T. aquaticus*	X-ray	N and G domains of Ffh with non-hydrolyzable GTP analog GMPPNP (N2 = crystal form 2)	0.190	1.9	Padmanabhan et al 2001[71]
1LS1	*T. aquaticus*	X-ray	Apo Ffh N and G domains	0.137	1.1	Ramirez et al 2002[72]
1QZX	*S. solfataricus*	X-ray	Complete Ffh without helix 8	0.313	4.0	Rosendal et al 2003[16]
1O87	*T. aquaticus*	X-ray	N and G domains of Ffh with MgGDP	0.197	2.1	Focia et al 2004[20]
1OKK	*T. aquaticus*	X-ray	N and G domains of Ffh in complex with N and G domain of FtsY	0.156	2.0	Focia et al 2004[20]
1RJ9	*T. aquaticus*	X-ray	N and G domains of Ffh in complex with N and G domain of FtsY	0.206	1.9	Egea et al 2004[21]
1FTS	*E. coli*	X-ray	FtsY	0.222	2.2	Montoya et al 1997[19]

Table 1C. SRP RNA structures

PDB ID	Source	Method	Description	R Value	Resolution [Å]	Reference
1CQ5	*E. coli*	NMR	SRP RNA domain IV (43-mer) (*averaged structure*)	n/a	n/a	Schmitz et al 1999[73]
1CQL	*E. coli*	NMR	SRP RNA domain IV (43-mer) (*ensemble*)	n/a	n/a	Schmitz et al 1999[73]
1DUH	*E. coli*	X-ray	The conserved domain IV of 4.5S RNA	0.230	2.7	Jovine et al 2000[74]
28SP	*E. coli*	NMR	The most conserved RNA motif in SRP RNA (SRP54 Binding Domain) (*ensemble*)	n/a	n/a	Schmitz et al In press
28SR	*E. coli*	NMR	The most conserved RNA motif in SRP RNA (SRP54 Binding Domain) (*averaged structure*)	n/a	n/a	Schmitz et al In press

direct interaction between FtsY and the translocation machinery, which then facilitates the insertion of the targeted protein into the SecYEG channel. The chloroplast FtsY forms a large complex that includes SecY and Alb3, the chloroplast YidC homolog.[23] A very active area of research addresses the mechanism by which membrane proteins enter the channel and translocate their hydrophilic regions across the membrane. What part of the SecYEG complex does the hydrophobic domain of the membrane protein bind to? How is the hydrophobic segment of the inserting membrane protein released from the Sec machinery and integrated into the bilayer? How does the Sec machinery perform these translocation and integration functions while maintaining a tight seal to prevent exchange of ions and solutes across the membrane?

In Bacteria, the Sec components SecY and SecE form the minimum translocation machinery.[24] SecYEG is sufficient to insert the membrane protein FtsQ in vitro.[25] Although SecDF is not essential for insertion, it does facilitate translocation.[26] SecG also promotes protein translocation but is not essential for insertion.[26-28] In some cases, SecA uses the energy of ATP hydrolysis to promote translocation of the hydrophilic domain of a membrane protein across the membrane.

The SecA-driven translocation of hydrophilic domains of membrane proteins most likely occurs in steps of 20 to 25 amino acid residues, as shown for the exported protein proOmpA.[29] By this same mechanism, the SecA bound to a membrane protein inserts into the membrane upon ATP binding, taking with it a segment (20 to 25 residues) of the polypeptide domain to be translocated. Following ATP hydrolysis, SecA dissociates from the membrane protein and returns to the cytoplasmic side of the membrane. By repeated cycles of SecA insertion and deinsertion, the polypeptide domain of the membrane protein is moved across the membrane.

Structure of SecA

SecA is a multifaceted protein. It binds to phospholipids, ATP, SecY, signal peptide, the mature domain of exported proteins, and SecB (for review, see ref 30). For membrane proteins, SecA is believed to bind to the hydrophobic domain (analogous to a signal peptide), and a part of the hydrophilic domain to be translocated. SecA belongs to the group of ATPases that show similarity to the DEAD-box helicases.[31,32]

Crystal structures are available for SecA from *Bacillus subtilis* (Fig. 3A) (Table 2)[33] and *Mycobacterium tuberculosis.*[34] These were solved both in the apo-form and in complex with ADP. Both studies were consistent with an antiparallel physiological dimer which was seen in solution by FRET experiments. Interestingly, the packing interactions are different in the two structures. *B. subtilis* SecA has also been crystallized under conditions that result in a monomeric form of the SecA (Fig. 3B) which adopts a more open conformation than the dimeric form.[35] Previous biochemical studies had shown that interaction with SecY, acidic phospholipids or signal peptides induces SecA into a monomeric form[36] with significant conformational changes. The monomeric crystals also gave improved resolution, diffracting to 2.2 Å resolution and revealing interpretable electron density for most of the molecule (Fig. 3B). The structure of SecA can be thought of as having two separate regions, the motor region and the translocation region. The motor region is made up of two nucleotide-binding fold domains (NBF1 and NBF2) and the translocation region is made up of the preprotein crosslinking domain (PPXD), the helical wing domain (HWD) and the helical scaffold domain (HSD). From the crystal structures, the binding of ADP does not appear to change the structures of the NBF domains. The major difference between the dimeric (Fig. 3C) and monomeric (Fig. 3B) forms of SecA is a result of an approximately 60° rotation of the PPXD and a rotation of the HWD and HSD of approximately 15°, resulting in the formation of a large groove between the PPXD, HSD and HWD. This groove is postulated to be the peptide-binding site. In all the crystal structures of the entire SecA protein available so far, there has not been experimental electron density for the C-terminal zinc-binding domain. The structure of this zinc-binding domain alone has been solved in solution by NMR.[37,38] It has also been solved by X-ray crystallography in complex with the targeting protein SecB.[39]

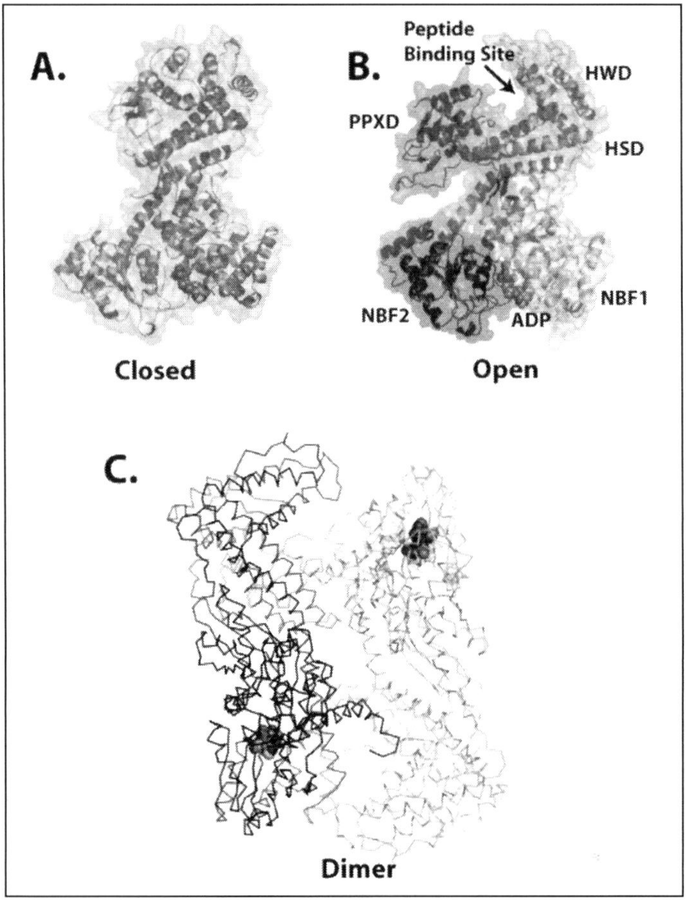

Figure 3. A) A ribbon diagram with transparent surface showing the closed form of SecA from *Bacillus subtilis*.[33] B) A ribbon diagram with transparent surface showing the open monomeric form of SecA from *Bacillus subtilis*.[35] The nucleotide-binding fold 1 domain (NBF1), the nucleotide binding fold 2 domain (NBF2), the helical scaffold domain (HSD), the preprotein cross-linking domain (PPXD) and the helical wing domain (HWD) are labeled. The bound ADP is shown in van der Waals spheres. The deep groove proposed to be the signal sequence binding domain is pointed out with an arrow. PDB coordinates 1TF5 and the program PyMol[66] were used to make this figure. C) A Cα trace diagram showing the dimeric closed state of the SecA protein from *Bacillus subtilis*.[33] The bound ADP is shown in van der Waals spheres. PDB coordinates 1M74 and the program PyMol[66] were used to make both Figure 3A,C.

Structure of the SecYEβ Complex

A central question in the membrane protein biogenesis and protein export field asks what are the structural and mechanistic characteristics of the protein-conducting SecYEG machinery. 2D electron microscopy studies provided the first clues to this question. The oligomeric forms of SecYEG complex from *E. coli*[40] and *B. subtilis*[41] are dimers. However, there are some tetramers that form when SecA is bound.[40] A three-dimensional structure of the *Escherichia coli* SecYEG complex was initially reported from cryo-electron microscopy analysis of 2D crystals.[42] The results suggested that SecYEG was a dimer with a closed cavity at the interface between the two monomers.

Table 2. SecA structures

PDB ID	Source	Method	Description	R Value	Resolution [Å]	Reference
1M6N	*B. subtilis*	X-ray	Nucleotide-free	0.220	2.7	Hunt et al 2002[33]
1M74	*B. subtilis*	X-ray	Mg-ADP-bound	0.217	3.0	Hunt et al 2002[33].
1NKT	*M. tuberculosis*	X-ray	Complex with ADP-βS	0.216	2.6	Sharma et al 2003[34].
1NL3	*M. tuberculosis*	X-ray	Nucleotide-free	0.196	2.8	Sharma et al 2003[34]
1TF2	*B. subtilis*	X-ray	Monomeric, open conformation, ADP-bound	0.228	2.9	Osborne et al 2004[35]
1TF5	*B. subtilis*	X-ray	Monomeric, open conformation, nucleotide-free	0.241	2.2	Osborne et al 2004[35]
1OZB	*H. influenzae*	X-ray	SecB complexed with the SecA C-terminus	0.226	2.8	Zhou and Xu, 2003[39]
1SX0	*E. coli*	NMR	C-terminal zinc-binding domain of SecA	n/a	n/a	Dempsey et al 2004[37]
1SX1	*E. coli*	NMR	C-terminal zinc-binding domain of SecA	n/a	n/a	Dempsey et al 2004[37]
1TM6	*E. coli*	NMR	The free zinc-binding C-terminal domain of SecA	n/a	n/a	Matousek and Alexandrescu, 2004[38]

Table 3. SecYEG structures

PDB ID	Source	Method	Description	R Value	Resolution [Å]	Reference
1RH5	*M. jannaschii*	X-ray	Double mutant	0.242	3.2	Van Den Berg et al 2004[47]
1RHZ	*M. jannaschii*	X-ray	Wild-type	0.254	3.5	Van Den Berg et al 2004[47]
a	*E. coli*	Cryo-EM		n/a	8.0	Breyton et al 2002[42]

a. The electron density file of a SecYEG dimer with the non-crystallographic symmetry imposed is available from the Supplementary Information from the Nature webpage (http://www.nature.com).

Low-resolution cryo-EM studies have also been performed on the ER Sec translocon. These studies revealed an oligomeric Sec61 complex[43] and a ribosome-Sec61 complex with the pore of the Sec61 complex aligning with the exit tunnel located within the large ribosomal subunit.[44] It was suggested that the central part of the Sec61 complex represented an aqueous pore because previous studies using fluorescently-labeled polypeptide chain positioned within the translocation channel suggested the chain to be in an aqueous environment.[45] Additionally, the size of the aqueous channel was determined to be 40 to 60 Å.[46]

Then came the big surprise in 2004 with the X-ray crystal structure of the SecY complex (SecYEβ) from Archaea[47] (Table 3). The crystal structure of the heterotrimeric SecYEβ complex was solved to 3.2 Å resolution in the presence of the detergent diheptanoylphosphatidyl choline (Fig. 4A). The archaeon *Methanococcus janaschii* was chosen as the source of the Sec components, based on the stability and crystallizability of the complex after screening proteins from 10 different species. The structure shows that the SecY (Sec61α subunit) protein consists of 10 transmembrane segments with the helices packed such that the protein makes two symmetrical halves with both the amino- and carboxy-termini facing the cytoplasm (transmembrane segments 1-5 and 6-10 form the symmetrical halves). The Sec61β and SecE(Sec61γ) subunits each have one transmembrane segment with the amino terminus facing the cytoplasm (Fig. 4A). Surprisingly, the structure suggests that the translocase pore resides at the center of one copy of the heterotrimeric SecYEβ. As mentioned above, previous biochemical and cryo-electron microscopy evidence had suggested that the pore may be assembled from multiple copies of SecYEβ. A cross-section of the channel reveals an overall shape of an hourglass with a ring of isoleucine residues that lines the constriction point (approximately 3 Å in diameter) near the center of the membrane. Interestingly, the structure of the channel, which is presumably in the closed state, reveals a small helix that sits on top of the pore and plugs the channel. The extremely small diameter of the pore suggests that the transmembrane segments of SecYEβ would go through a significant rearrangement in the open state to accommodate a substrates in the process of translocation. Further conformational changes would be needed for the α-helix of the inserting membrane protein to escape the channel and partition into the lipids of the membrane. The authors propose a ribosomal binding surface for the homologous eukaryotic translocon, and a binding site for the SecA ATPase in Eubacteria.[47] There is approximately 50% sequence similarity in the eubacterial and eukaryotic genes SecYE and Sec61αγ, respectively. Sec61β and SecG show no sequence similarity.

A model can be proposed, based on the structure of the SecYEβ complex and biochemical data, on how the hydrophobic transmembrane helix of a membrane protein binds to the SecYEβ complex.[47] The helix would bind to the SecYEβ complex in a manner analogous to how binding of a signal peptide of an exported protein was proposed.[47] Binding would cause

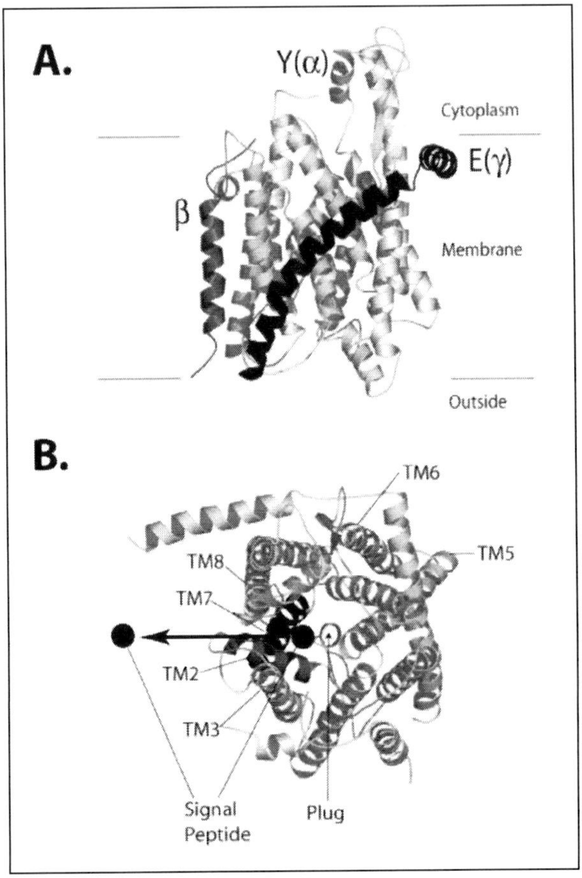

Figure 4. A) A ribbon diagram showing the heterotrimeric complex of SecYEβ from the archaeon *Methanococcus jannaschii*.[47] The SecY (or Sec61α subunit), SecE (or Sec61γ subunit) and Sec61β subunit (no sequence similarity to SecG) are labeled. The orientation of the channel is shown relative to the phospholipid bilayer it resides in. PDB coordinates 1RHZ and the program PyMol[66] were used to make this figure. B) A ribbon diagram showing the top (cytoplasmic) view of the SecYEβ channel (light grey). The transmembrane segments TM2 and TM7, proposed to be part of the exit route for the substrates hydrophobic segments, are rendered in a dark shade and labeled. The short helical plug in the center of the channel is also labeled. A black dot designates the possible positions of a signal peptide.

dissociation of the plug from the pore, thereby allowing initiation of the steps of translocation to proceed (Fig. 4B). In the case of a membrane protein containing one hydrophobic transmembrane helix, binding would allow the hydrophilic flanking region to pass through the channel in a manner that needs to be defined in the future. For a membrane protein that is cotranslationally inserted into the membrane, the ribosome is most likely bound to the SecY complex and the energy driving translocation is derived from protein synthesis (Fig. 5B). For SecA-dependent translocation of the hydrophilic domain, the ribosome of the nascent membrane protein complex would have to detach from SecYEG in order for SecA to bind to SecYEG and initiate translocation of the hydrophilic region in steps of 20 to 25 residues (Fig. 5C). How this is achieved is not clear. SecA could lead to translocation of the polypeptide chain by a region of SecA itself moving through the channel. However, it is hard

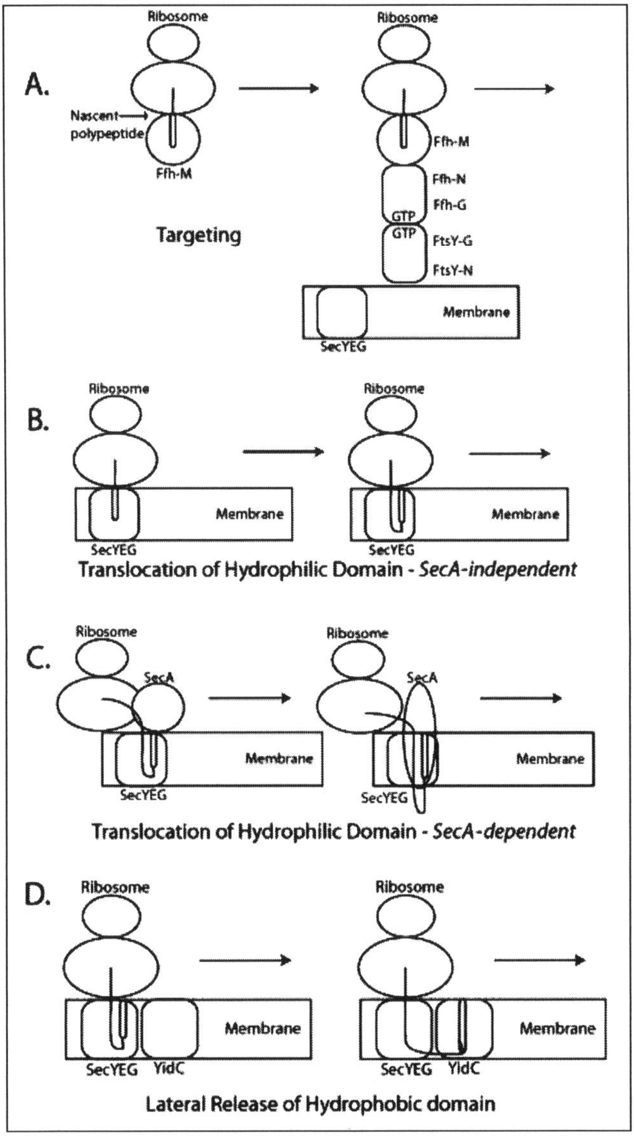

Figure 5. A schematic depiction of the possible individual steps of membrane protein assembly. A) Targeting of the nascent protein to the membrane. The Ffh-bound nascent chain is targeted to the membrane in a GTP-dependent manner by the interaction of the Ffh NG domain with the NG domain of FtsY. B) SecA-independent translocation of a hydrophobic domain. Translocation of the chain within the channel is driven by the energy of protein synthesis. C) SecA-dependent translocation of a hydrophilic domain. The binding of SecA to the protein chain drives translocation of a loop across the membrane. D) Release of the hydrophobic domain from the SecYEG complex. After release of the hydrophobic segment from the SecYEG channel, the transmembrane segment is stabilized by YidC. See the text for details of the individual steps.

to imagine how this could occur with a monomeric SecYEG complex. Or SecA itself does not penetrate the Sec complex. Alternatively, SecA could simply bind to the SecYEG channel, thereby causing a conformational change in the SecY complex that opens the channel to allow translocation of the polypeptide chain.

Lateral Integration, Assembly and Folding

Recent studies have focused on how membrane proteins laterally integrate into the membrane bilayer after inserting into the SecY complex and then assemble into their three-dimensional structure. Van der Berg et al[47] hypothesized that the substrate's hydrophobic transmembrane helices may escape from the channel via the interface between the two symmetrical halves of the SecY protein. The structural information, along with previous photocrosslinking data,[48] suggests that newly assembling transmembrane domains (anchor segment) of membrane proteins may insert between SecY transmembrane segments TM7 and TM2 (Fig. 4B) which make up a lateral gate (along with TM8 and TM3) through which the newly assembling transmembrane segments may partition into the surrounding lipid. The insertion between TM7 and part of TM2 would also trigger an opening of the channel structure allowed by a proposed ~15° hinge motion between TM5 and TM6 (the connection point between the two pseudo-symmetrically related halves of the SecY molecule). The hinge motion of the structure would allow for a proposed 15-20Å by 10-15Å pore opening for the insertion of the anchor segment loop. However, the process may be mediated by YidC specifically recognizing transmembrane regions of membrane proteins in *E. coli.*[7,49] YidC has been suggested to function as an assembly site for hydrophobic regions of mannitol permease (MtlA). Muller and coworkers showed that hydrophobic domain 3 of a nascent Mtl membrane protein inserts at the SecY/YidC interface while the hydrophobic domain 1 and 2 are still in contact with YidC.[50] Therefore, after the hydrophobic region leaves laterally from the SecYEG complex it may interact with YidC which would stabilize the hydrophobic region until it integrates into the membrane (Fig. 5D).

Even more recently, the best evidence thus far for YidC playing a role in folding of a membrane protein was presented. Nagamori et al showed that lac permease, which spans the membrane twelve times, inserts quite normally when membranes contain deficient levels of YidC.[51] However, the inserted lac permease under YidC-depleted conditions appears to be aberrantly folded as monoclonal antibodies that specifically recognize certain periplasmic loops of lac permease are impaired in their binding.

The role of YidC in the insertion of Sec-dependent proteins varies, depending on the membrane protein being studied. For membrane proteins such as Lep and FtsQ, which have large C-terminal domains, YidC does not play an important translocation role.[52,53] However, YidC is required in vivo for insertion of the Sec-dependent a and b subunits of the F_1F_0ATP synthase.[54]

Presently, effort is needed to solve the structure of YidC so as to reveal key features of the protein such as whether YidC has channel or transporter properties. To provide information about the region of YidC important for its membrane insertase function, we have studied a detailed collection of deletion and substitution mutants.[55] YidC is a 60 kDa integral membrane protein with six transmembrane segments. Transmembrane regions two, three and six are important for activity and contain residues that are critical for membrane insertase activity. It will be necessary to determine which parts of YidC constitute the substrate-binding region and how the transmembrane segments within the protein interact. In addition, the oligomeric structure of YidC will also need to be determined within intact membranes. The formation of a structure of YidC within intact membrane would explain why some purified YidC appears as a dimer upon blue native polyacrylamide electrophoresis.[56] It should be noted that Oxa1, the mitochondrial homolog of YidC, is a tetramer.[57]

Insertion by the Novel YidC Pathway

The second route by which proteins can insert into the membrane is by the YidC pathway (for review, see ref. 58) (Fig. 1B). Strong evidence for the evolutionarily-conserved nature of this insertion pathway was obtained when it was discovered that the Sec-independent phage M13 procoat and Pf3 coat proteins require YidC for membrane insertion.[52,59] Previously, the YidC homologs in mitochondria and chloroplasts (Oxa1 and Alb3, respectively) were found to play a role in membrane protein insertion in these organelles (for review, see ref. 60). YidC plays a direct role in the membrane insertion process as it comes into contact with Pf3 coat protein during membrane insertion of the phage protein.[49] Both M13 procoat and Pf3 coat protein do not require the SRP pathway for insertion.

To date, the only endogenous *E. coli* protein that has been discovered to require YidC and insert by a Sec-independent mechanism is subunit c of the F_1F_0ATP synthase.[54,61] In vivo and in vitro studies have demonstrated that subunit c inserts independent of the Sec translocase and does not require the SRP targeting components for insertion (see ref. 62 for a differing opinion). Unlike the M13 procoat and Pf3 coat proteins, insertion of subunit c does not require the proton motive force across the membrane. Yet, like the M13 procoat and Pf3 coat proteins, subunit c is small in size and has short translocated regions. What structural features render a protein completely dependent on YidC for its membrane insertion are not known.

It is not clear whether YidC acts alone in intact cells or whether there are other proteins which make membrane insertion more efficient or regulate insertion by this pathway. Interestingly, the mitochondrial homolog Oxa1 has recently been shown to bind to the ribosome,[63,64] specifically to the large ribosomal protein Mrp20 (homologous to the L23 *E. coli* protein).[63] This raises the question of whether YidC also binds to the ribosome in *E. coli*.

Conclusions and Future Questions

During the last few years, we have seen the first three-dimensional structures of the membrane-localized protein-conducting channel, its ATPase motor, and the SRP targeting components Ffh and FtsY. These structures have provided tremendous insight into the role these proteins play in membrane protein biogenesis. However, there are limitations to the current work because they do not provide information on the dynamic nature of the components during the translocation process. Despite significant advances in this area, we are only now beginning to understand how membrane proteins are assembled within the lipid bilayer.

To understand how the membrane protein inserts into the lipid bilayer and folds into a stable and active conformation, it will be necessary to shed light on how the protein partitions into the membrane, where the helical transmembrane segments associate to form the transmembrane domain of the protein. This will require a multidisciplinary approach involving biophysical studies, structural analysis as well as cell biology, genetics and molecular biology.

The recent availability of the three-dimensional structures of the proteins and protein complexes involved in membrane assembly now provides the opportunity for detailed structure-function studies and for molecular dynamics simulation analysis which could provide important insights into the mechanism of this very dynamic molecular machinery. Cryo-electron microscopy experiments with 2D crystals will also be helpful in the understanding of the movements within this system.

To provide a deeper understanding of the membrane insertion mechanism, it will be necessary to examine how membrane proteins interact with the Sec machinery at an atomic level. Thus, future directions include determining the three-dimensional structure of Sec complex intermediates, such as the SecYEG complex with a bound membrane protein, signal peptide or in complex with SecA as well as with the channel in the open state. Elucidating the structures of the eubacterial SecDFyajC complex and YidC may help to provide clues as to their functional roles in membrane protein translocation and membrane protein assembly. Also needed are high-resolution structures of a signal peptide bound to the M domain of Ffh of the SRP and the complete Ffh-FtsY structure. Other open questions remaining in this field which will require biochemical, biophysical and genetic methodologies to answer are: How are membrane proteins integrated into the lipid bilayer after they leave the Sec complex and what is the role of YidC in this process? Does YidC have a general chaperone-like function to help fold membrane proteins? Answering these questions is essential to obtaining a complete picture of how proteins are assembled into the membrane.

References

1. von Heijne G. Recent advances in the understanding of membrane protein assembly and structure. Q Rev Biophys 1999; 32:285-307.
2. Aridor M, Hannan LA. Traffic jam: A compendium of human diseases that affect intracellular transport processes. Traffic 2000; 1:836-851.
3. Aridor M, Hannan LA. Traffic jams II: An update of diseases of intracellular transport. Traffic 2002; 3:781-790.
4. Koch HG, Moser M, Muller M. Signal recognition particle-dependent protein targeting, universal to all kingdoms of life. Rev Physiol Biochem Pharmacol 2003; 146:55-94.
5. Keenan RJ, Freymann DM, Stroud RM et al. The signal recognition particle. Annu Rev Biochem 2001; 70:755-775.
6. Driessen AJ, Manting EH, van der Does C. The structural basis of protein targeting and translocation in bacteria. Nat Struct Biol 2001; 8:492-498.
7. Scotti PA, Urbanus ML, Brunner J et al. YidC, the Escherichia coli homologue of mitochondrial Oxa1p, is a component of the Sec translocase. EMBO J 2000; 19:542-549.
8. Kuhn A. Alterations in the extracellular domain of M13 procoat protein make its membrane insertion dependent on secA and secY. Eur J Biochem 1988; 177:267-271.
9. Andersson H, von Heijne G. Sec dependent and sec independent assembly of E. coli inner membrane proteins: The topological rules depend on chain length. EMBO J 1993; 12:683-691.
10. Neumann-Haefelin C, Schafer U, Muller M et al. SRP-dependent cotranslational targeting and SecA-dependent translocation analyzed as individual steps in the export of a bacterial protein. EMBO J 2000; 19:6419-6426.
11. Romisch K, Webb J, Herz J et al. Homology of 54K protein of signal-recognition particle, docking protein and two E. coli proteins with putative GTP-binding domains. Nature 1989; 340:478-482.
12. Lee HC, Bernstein HD. The targeting pathway of Escherichia coli presecretory and integral membrane proteins is specified by the hydrophobicity of the targeting signal. Proc Natl Acad Sci USA 2001; 98:3471-3476.
13. Phillips GJ, Silhavy TJ. The E. coli ffh gene is necessary for viability and efficient protein export. Nature 1992; 359:744-746.
14. Luirink J, ten Hagen-Jongman CM, van der Weijden CC et al. An alternative protein targeting pathway in Escherichia coli: Studies on the role of FtsY. EMBO J 1994; 13:2289-2296.
15. Miller JD, Bernstein HD, Walter P. Interaction of E. coli Ffh/4.5S ribonucleoprotein and FtsY mimics that of mammalian signal recognition particle and its receptor. Nature 1994; 367:657-659.
16. Rosendal KR, Wild K, Montoya G et al. Crystal structure of the complete core of archaeal signal recognition particle and implications for interdomain communication. Proc Natl Acad Sci USA 2003; 100:14701-14706.
17. Keenan RJ, Freymann DM, Walter P et al. Crystal structure of the signal sequence binding subunit of the signal recognition particle. Cell 1998; 94:181-191.
18. Batey RT, Rambo RP, Lucast L et al. Crystal structure of the ribonucleoprotein core of the signal recognition particle. Science 2000; 287:1232-1239.
19. Montoya G, Svensson C, Luirink J et al. Crystal structure of the NG domain from the signal-recognition particle receptor FtsY. Nature 1997; 385:365-368.
20. Focia PJ, Shepotinovskaya IV, Seidler JA et al. Heterodimeric GTPase core of the SRP targeting complex. Science 2004; 303:373-377.
21. Egea PF, Shan SO, Napetschnig J et al. Substrate twinning activates the signal recognition particle and its receptor. Nature 2004; 427:215-221.
22. Shan SO, Walter P. Induced nucleotide specificity in a GTPase. Proc Natl Acad Sci USA 2003; 100:4480-4485.

23. Moore M, Goforth RL, Mori H et al. Functional interaction of chloroplast SRP/FtsY with the ALB3 translocase in thylakoids: Substrate not required. J Cell Biol 2003; 162:1245-1254.
24. Akimaru J, Matsuyama S, Tokuda H et al. Reconstitution of a protein translocation system containing purified SecY, SecE, and SecA from Escherichia coli. Proc Natl Acad Sci USA 1991; 88:6545-6549.
25. van der Laan M, Nouwen N, Driessen AJ. SecYEG proteoliposomes catalyze the Deltaphi-dependent membrane insertion of FtsQ. J Biol Chem 2004; 279:1659-1664.
26. Duong F, Wickner W. Distinct catalytic roles of the SecYE, SecG and SecDFyajC subunits of preprotein translocase holoenzyme. EMBO J 1997; 16:2756-2768.
27. Nishiyama K, Suzuki T, Tokuda H. Inversion of the membrane topology of SecG coupled with SecA-dependent preprotein translocation. Cell 1996; 85:71-81.
28. Matsumoto G, Mori H, Ito K. Roles of SecG in ATP- and SecA-dependent protein translocation. Proc Natl Acad Sci USA 1998; 95:13567-13572.
29. Economou A, Wickner W. SecA promotes preprotein translocation by undergoing ATP-driven cycles of membrane insertion and deinsertion. Cell 1994; 78:835-843.
30. Vrontou E, Karamanou S, Baud C et al. Global coordination of protein translocation by the SecA IRA1 switch. J Biol Chem 2004; 279:22490-22497.
31. Subramanya HS, Bird LE, Brannigan JA et al. Crystal structure of a DExx box DNA helicase. Nature 1996; 384:379-383.
32. Kim JL, Morgenstern KA, Griffith JP et al. Hepatitis C virus NS3 RNA helicase domain with a bound oligonucleotide: The crystal structure provides insights into the mode of unwinding. Structure 1998; 6:89-100.
33. Hunt JF, Weinkauf S, Henry L et al. Nucleotide control of interdomain interactions in the conformational reaction cycle of SecA. Science 2002; 297:2018-2026.
34. Sharma V, Arockiasamy A, Ronning DR et al. Crystal structure of Mycobacterium tuberculosis SecA, a preprotein translocating ATPase. Proc Natl Acad Sci USA 2003; 100:2243-2248.
35. Osborne AR, Clemons Jr WM, Rapoport TA. A large conformational change of the translocation ATPase SecA. Proc Natl Acad Sci USA 2004; 101:10937-10942.
36. Or E, Navon A, Rapoport T. Dissociation of the dimeric SecA ATPase during protein translocation across the bacterial membrane. EMBO J 2002; 21:4470-4479.
37. Dempsey BR, Wrona M, Moulin JM et al. Solution NMR structure and X-ray absorption analysis of the C-terminal zinc-binding domain of the SecA ATPase. Biochemistry 2004; 43:9361-9371.
38. Matousek WM, Alexandrescu AT. NMR structure of the C-terminal domain of SecA in the free state. Biochim Biophys Acta 2004; 1702:163-171.
39. Zhou J, Xu Z. Structural determinants of SecB recognition by SecA in bacterial protein translocation. Nat Struct Biol 2003; 10:942-947.
40. Manting EH, van Der Does C, Remigy H et al. SecYEG assembles into a tetramer to form the active protein translocation channel. EMBO Journal 2000; 19:852-861.
41. Meyer TH, Menetret JF, Breitling R et al. The bacterial SecY/E translocation complex forms channel-like structures similar to those of the eukaryotic Sec61p complex. J Mol Biol 1999; 285:1789-1800.
42. Breyton C, Haase W, Rapoport TA et al. Three-dimensional structure of the bacterial protein-translocation complex SecYEG. Nature 2002; 418:662-665.
43. Hanein D, Matlack KE, Jungnickel B et al. Oligomeric rings of the Sec61p complex induced by ligands required for protein translocation. Cell 1996; 87:721-732.
44. Beckmann R, Bubeck D, Grassucci R et al. Alignment of conduits for the nascent polypeptide chain in the ribosome-Sec61 complex. Science 1997; 278:2123-2126.
45. Crowley KS, Liao S, Worrell VE et al. Secretory proteins move through the endoplasmic reticulum membrane via an aqueous, gated pore. Cell 1994; 78:461-471.
46. Hamman BD, Chen JC, Johnson EE et al. The aqueous pore through the translocon has a diameter of 40-60 A during cotranslational protein translocation at the ER membrane. Cell 1997; 89:535-544.
47. Van den Berg B, Clemons Jr WM, Collinson I et al. X-ray structure of a protein-conducting channel. Nature 2004; 427:36-44.
48. Plath K, Mothes W, Wilkinson BM et al. Signal sequence recognition in posttranslational protein transport across the yeast ER membrane. Cell 1998; 94:795-807.
49. Chen M, Samuelson JC, Jiang F et al. Direct interaction of YidC with the Sec-independent Pf3 coat protein during its membrane protein insertion. J Biol Chem 2002; 277:7670-7675.

50. Beck K, Eisner G, Trescher D et al. YidC, an assembly site for polytopic Escherichia coli membrane proteins located in immediate proximity to the SecYE translocon and lipids. EMBO Rep 2001; 2:709-714.
51. Nagamori S, Smirnova IN, Kaback HR. Role of YidC in folding of polytopic membrane proteins. J Cell Biol 2004; 165:53-62.
52. Samuelson JC, Chen M, Jiang F et al. YidC mediates membrane protein insertion in bacteria. Nature 2000; 406:637-641.
53. Urbanus ML, Scotti PA, Froderberg L et al. Sec-dependent membrane protein insertion: Sequential interaction of nascent FtsQ with SecY and YidC. EMBO Rep 2001; 2:524-529.
54. Yi L, Celebi N, Chen M et al. Sec/SRP requirements and energetics of membrane insertion of subunits a, b, and c of the Escherichia coli F1F0 ATP synthase. J Biol Chem 2004; 279:39260-39267.
55. Jiang F, Chen M, Yi L et al. Defining the regions of Escherichia coli YidC that contribute to activity. J Biol Chem 2003; 278:48965-48972.
56. Nouwen N, Driessen AJ. SecDFyajC forms a heterotetrameric complex with YidC. Mol Microbiol 2002; 44:1397-1405.
57. Nargang FE, Preuss M, Neupert W et al. The Oxa1 protein forms a homooligomeric complex and is an essential part of the mitochondrial export translocase in Neurospora crassa. J Biol Chem 2002; 277:12846-12853.
58. Dalbey RE, Kuhn A. YidC family members are involved in the membrane insertion, lateral integration, folding, and assembly of membrane proteins. J Cell Biol 2004; 166:769-774.
59. Samuelson JC, Jiang F, Yi L et al. Function of YidC for the insertion of M13 procoat protein in E. coli: Translocation of mutants that show differences in their membrane potential dependence and Sec- requirement. J Biol Chem 2001; 16:16.
60. Kuhn A, Stuart R, Henry R et al. The Alb3/Oxa1/YidC protein family: Membrane-localized chaperones facilitating membrane protein insertion? Trends Cell Biol 2003; 13:510-516.
61. Van Der Laan M, Bechtluft P, Kol S et al. F1F0 ATP synthase subunit c is a substrate of the novel YidC pathway for membrane protein biogenesis. J Cell Biol 2004; 165:213-222.
62. van Bloois E, Jan Haan G, de Gier JW et al. F(1)F(0) ATP synthase subunit c is targeted by the SRP to YidC in the E. coli inner membrane. FEBS Lett 2004; 576:97-100.
63. Jia L, Dienhart M, Schramp M et al. Yeast Oxa1 interacts with mitochondrial ribosomes: The importance of the C-terminal region of Oxa1. EMBO J 2003; 22:6438-6447.
64. Szyrach G, Ott M, Bonnefoy N et al. Ribosome binding to the Oxa1 complex facilitates cotranslational protein insertion in mitochondria. EMBO J 2003; 22:6448-6457.
65. Harms J, Schluenzen F, Zarivach R et al. High resolution structure of the large ribosomal subunit from a mesophilic eubacterium. Cell 2001; 107:679-688.
66. The PyMOL Molecular Graphics System [computer program]. Version 0.96. San Carlos, CA, USA: DeLano Scientific, 2002.
67. Batey RT, Sagar MB, Doudna JA. Structural and energetic analysis of RNA recognition by a universally conserved protein from the signal recognition particle. J Mol Biol 2001; 307:229-246.
68. Freymann DM, Keenan RJ, Stroud RM et al. Structure of the conserved GTPase domain of the signal recognition particle. Nature 1997; 385:361-364.
69. Freymann DM, Keenan RJ, Stroud RM et al. Functional changes in the structure of the SRP GTPase on binding GDP and Mg2+GDP. Nat Struct Biol 1999; 6:793-801.
70. Montoya G, Kaat K, Moll R et al. The crystal structure of the conserved GTPase of SRP54 from the archaeon Acidianus ambivalens and its comparison with related structures suggests a model for the SRP-SRP receptor complex. Structure Fold Des 2000; 8:515-525.
71. Padmanabhan S, Freymann DM. The conformation of bound GMPPNP suggests a mechanism for gating the active site of the SRP GTPase. Structure (Camb) 2001; 9:859-867.
72. Ramirez UD, Minasov G, Focia PJ et al. Structural basis for mobility in the 1.1 A crystal structure of the NG domain of Thermus aquaticus Ffh. J Mol Biol 2002; 320:783-799.
73. Schmitz U, Behrens S, Freymann DM et al. Structure of the phylogenetically most conserved domain of SRP RNA. RNA 1999; 5:1419-1429.
74. Jovine L, Hainzl T, Oubridge C et al. Crystal structure of the ffh and EF-G binding sites in the conserved domain IV of Escherichia coli 4.5S RNA. Structure Fold Des 2000; 8:527-540.

Chapter 6

The Twin-Arginine Transport System

Frank Sargent, Ben C. Berks and Tracy Palmer*

Abstract

The twin-arginine transport (Tat) system is a protein-targeting pathway found in the cytoplasmic membranes of many eubacteria, some Archaea, and the chloroplasts and mitochondria of plants. It is apparently not a feature of animal physiology. Substrate proteins are targeted to a membrane-bound transport apparatus by N-terminal signal peptides harbouring a distinctive 'twin-arginine' amino acid sequence motif, and, most remarkably, all substrate proteins are transported in a fully folded conformation. Model systems most commonly used to study the fundamentals of Tat transport are the Gram-negative eubacterium *Escherichia coli*, the Gram-positive eubacterium *Bacillus subtilis*, and thylakoid membranes derived from pea or maize chloroplasts. Here, we have attempted to integrate our knowledge of the key aspects of these well-characterized Tat protein transport pathways, to carve-out some shared principles between systems, and arrive at a broad consensus covering the physiology and biochemistry of Tat transport.

Traffic on the Tat Pathway

The majority of substrates that are exported by the bacterial Tat pathway are proteins containing redox-active cofactors.[1] Strong evidence suggests that such proteins acquire their cofactors, and therefore attain a folded conformation, in the cytosol prior to export (reviewed in ref. 2). Indeed, cofactor binding is usually taken as a prerequisite for Tat export since some *E. coli* enzymes are not exported before this process is complete.[3,4] It should be noted, however, that export of some Tat-dependent enzymes has been observed even when cofactor-insertion is blocked[5] and there are ever-increasing reports of bacterial Tat substrates that are devoid of prosthetic groups.[6-9] The existence of cofactor-lacking, but still fully folded, Tat substrates serves to highlight that transport of folded proteins is the raison d'etre of the Tat transporter. Conversely, there is now very strong evidence that the bacterial Tat system is incapable of transporting unfolded proteins.[10] These elegant experiments utilized *E. coli* alkaline phosphatase (PhoA), a protein that requires the formation of two disulfide bonds for correct folding and activity, as a Tat substrate. Tat-targeted PhoA was only transported when expressed in a mutant strain with an oxidizing, rather than the normal reducing, cytoplasm. This indicated that PhoA transport requires disulfide bond formation, and thus protein folding. As such, the Tat transporter exerts a quality control mechanism that rejects unfolded substrates.[10] A list of the 27

*Tracy Palmer—Department of Molecular Microbiology, John Innes Centre, Norwich NR4 7UH, U.K. Email: tracy.palmer@bbsrc.ac.uk

Table 1. *Escherichia coli* **Tat substrates grouped according to cofactor type**

Protein	Function
Iron Sulfur clusters	
• HyaA	[NiFe] hydrogenase-1 subunit
• HybO	[NiFe] hydrogenase-2 subunit
• HybA	electron transfer from hydrogenase-2
• NapG	electron transfer to nitrate reductase
• NrfC	electron transfer to nitrite reductase
• YagT	subunit of a Mo-dependent enzyme?
• YdhX	NrfC homolog?
Molybdopterin Guanine Dinucleotide (MGD)	
• TorA	TMAO reductase catalytic subunit
• TorZ	TMAO reductase-2 catalytic subunit
MGD and Iron Sulfur clusters	
• NapA	nitrate reductase catalytic subunit
• DmsA	DMSO reductase catalytic subunit
• YnfE	DMSO reductase homolog
• YnfF	DMSO reductase homolog
• FdnG	formate dehydrogenase-N catalytic subunit
• FdoG	formate dehydrogenase-O catalytic subunit
Molybdopterin	
• YedY	unknown
Plastocyanin-related proteins	
• CueO	multi-copper oxidase
• SufI	homologous to CueO [Cu ligands absent]
Other / no obvious cofactor	
• YahJ	Fe-dependent hydrolase?
• WcaM	biosynthesis of colanic acid
• YdcG	glucans biosynthesis?
• YcdB	unknown
• YcdO	unknown
• YaeI	phosphodiesterase?
• AmiA	cell wall amidase
• AmiC	cell wall amidase
• FhuD	ferrichrome binding protein

known or predicted Tat substrates from *E. coli* (which has a complete proteome of over 4,000 polypeptides) is given in Table 1.

The Twin-Arginine Signal Peptide

Proteins destined for export by the Tat pathway are synthesized with specialized N-terminal signal peptides bearing a conserved SRRxFLK 'twin-arginine' motif. The arginine side-chains are essentially invariant[6] although a very low number of signal peptides that disobey this rule, by the conserved substitution of one arginine by lysine, have been discovered in both prokaryotes and plants.[11,12] A number of site-directed mutagenesis experiments have confirmed the central importance of the arginine residues: For example, a complete export block was observed when the twin-arginine residues of the signal peptide of the *Wolinella succinogenes* [NiFe] hydrogenase were replaced with twin glutamines,[13] and transport of the glucose-fructose oxidoreductase (GFOR) from *Zymomonas mobilis* was prevented when the arginines were mutated to lysines.[14]

One of the most heavily exploited bacterial twin arginine signal peptides is that of the trimethylamine *N*-oxide (TMAO) reductase (TorA) from *E. coli*. The TorA signal peptide has been attached to green fluorescent protein (GFP),[15-17] Colicin V,[18] PhoA,[10] OEC23,[19] LepB,[20] MalE,[21] dimethyl sulfoxide reductase,[22] hydrogenase-2,[23] and GFOR.[24] Interestingly, the nature of the passenger protein to which the TorA signal is attached seems to have an influence on the operation of the Tat motif at the molecular level. For example, a mutant SR*K*RFLA TorA Tat motif was incapable of transporting the native TMAO reductase to the periplasm,[25] however TorA signals carrying this very sequence still directed export of both GFP[17] and Colicin V.[18] This is a quite remarkable, important and puzzling phenomenon, and the mechanism by which passenger proteins exert influence on signal peptide activity remains unknown.

Table 2. Tat signals from* E. coli *molybdenum-containing proteins

Protein	Enzyme	N-Terminal Signal Peptide
TorA	TMAO reductase	MNNNDLFQA **SRR**R**FL**AQ*LGGLTVAGMLGPSLL*T**P**RRATA
TorZ	TMAO reductase-2	MIREEVMTL **TRR**E**F**I**K**HS*GIAAGALVVTSAAPL***P**AWA
DmsA	DMSO reductase	MKTKIPDAVLAAEV **SRR**GLV**K**TT*AIGGLAMASSALTL***P**FSRIAHA
YnfE	DMSO reductase-2	MSKNERMVGI **SRR**TLV**K**ST*AIGSLALAAGGFSL***P**FTLRNAAA
YnfF	DMSO reductase-3	MKIHTTEALMKAEI **SRR**SLM**K**TS*ALGSLALASSAFTL***P**FSQMVRA
NapA	nitrate reductase	MKL **SRR**S**F**M**K**AN*AVAAAAAAAGLSV***P**GVARAVVG
FdnG	formate dehydrogenase-N	MDV **SRR**Q**F**F**K***ICAGGMAGTTVAALGFA***P**KQALA
FdoG	formate dehydrogenase-O	MQV **SRR**Q**F**F**K***ICAGGMAGTTAAALGFA***P**SVALA
YagT	?	MSNQGEYPEDNRVGKHEPHDLSL **TRR**DLI**K***VSAATAATAVVY***P**HSTLAASV
YedY	?	MKKNQFLKESDVTAESVFFM K**RR**QV**LK***ALGISATALSL***P**HAAHA

Residues contributing to the twin-arginine motif are bold underlined, side-chains of the signal peptide h-regions are italicized, conserved proline residues are shown in bold, and the signal peptide n-regions of variable sequence are boxed.

Outwith the twin-arginine motif, Tat signal peptides have a few other distinguishing features. N-terminal to the Tat motif lies the signal peptide 'n-region' that is very variable in sequence and normally polar in character. C-terminal to the Tat motif there always exists a hydrophobic 'h-region' of 15-25 amino acids that is often punctuated by a conserved proline residue preceeding a short polar 'c-region' which completes the signal peptide.[1,20] The majority of prokaryotic and plant twin-arginine signal peptides are proteolytically cleaved following the transport event, the exception being the Rieske Fe-S protein in which the signal peptide h-region ultimately forms a transmembrane helix through the lipid bilayer.[12]

To date, the functional importance of the signal peptide n-regions has not been extensively explored. It is notable that signal peptides for proteins binding similar cofactors (even those from different biological systems) exhibit striking sequence conservation in addition to the twin arginine motif (for example, the [NiFe] hydrogenase signal peptides aligned in Berks et al[26]). Interestingly, signal peptide sequence conservation between related Tat substrates is often located in the signal peptide n-region, and this region is also the most variable in sequence and length between unrelated passenger proteins (Table 2).

Ante-Transport Events

Tat transport is not just a post-translational event, it is a crucial biosynthetic step in assembly of a Tat substrate. All Tat substrates must at least be fully folded before Tat transport is attempted. In the case of bacterial redox enzymes, it is very important that Tat transport does not take place before cofactor-insertion is complete, and in some cases, before a signal-less partner protein that must be 'piggy-backed' through the Tat translocase has docked. For these reasons a system of 'proofreading' Tat transport has been postulated since the Tat translocase was first described.[3] The phrase 'proofreading' in the context of Tat transport is taken to mean the monitoring or checking of the state of assembly of a Tat substrate before export. It is fair to say that the phrase 'quality control' would have been the preferred description of this process. However, 'quality control' has been used to describe a mechanism by which the Tat apparatus itself accepts or rejects substrates presented for transport.[10] 'Proofreading' therefore describes early ante-transport events where Tat substrates are being prepared for export while 'quality control' describes a late ante-transport event where substrates 'pass or fail' at the final checkpoint. Tat transportation is the point of no return. Thus, for certain complex Tat substrates, the proofreading system is very important.

There is some evidence that twin-arginine signal peptides show specificity towards their native passenger proteins and their native host organism. GFOR, for example, is efficiently exported in the *Z. mobilis* bacterium but not when heterologously expressed in *E. coli*. Replacement of the GFOR Tat signal with that of *E. coli* TorA, however, allows transport of the chimera in *E. coli*.[24] Similarly, the chloroplast OEC23 protein was not exported when expressed in *E. coli*[27] unless the TorA signal was substituted for the native Tat signal peptide.[19] These observations suggest that Tat signal peptides are not universally recognized by different Tat translocases. However, to complicate the picture further, it should be noted that Tat signals from *Desulfovibrio vulgaris* are active in *E. coli*.[28]

There are also some indications that Tat signal peptides may be specific for particular enzymes within the same organism. For example, replacement of the signal peptide of *E. coli* DmsA by that of *E. coli* TorA impaired Tat transport of the modified DMSO reductase.[22] The molecular basis for this observation was originally not at all clear. However, a recent study of the Tat-dependent [NiFe] hydrogenase-2 isoenzyme may provide some belated explanation. As with the DMSO reductase experiment, assembly and activity of the hydrogenase-2 were severely impaired when the native signal peptide was swapped for that of TorA. However, this loss of function phenotype could be almost completely compensated for by overproduction of the TorD protein.[23] This suggests that TorD interacts with the TorA signal peptide. TorD is a 200 amino acid cytoplasmic protein that had previously been shown to be required for the efficient insertion of molybdopterin into TorA. Since cofactor insertion into TorA is required before Tat transport can proceed,[3] the intimate link between cofactor insertion and Tat transport provided by TorD might have been spotted earlier. However, since the TorA signal peptide is not essential for cofactor loading,[29] TorD was not widely expected to bind a twin-arginine signal peptide. Two-hybrid experiments confirmed that TorD binds directly to the TorA signal peptide,[23] while genetic and biochemical experiments confirmed a second, as-yet uncharacterized, binding site somewhere on the TorA enzyme itself.[23,30] Genetic experiments implicate the TorA signal peptide n-region as a recognition factor for TorD.[23] This would make some sense, as the twin-arginine motif itself is so highly conserved between all Tat substrates that this is unlikely to be the principle feature that a very specialized chaperone like TorD would recognize. The TorA signal peptide n-region is unique to TorA and, therefore, may help TorD to specifically recognize the TorA signal peptide. The binding of the TorA precursor by TorD seems to prevent premature targeting until all other assembly processes are complete—the very definition of an archetypal 'proofreading chaperone' on the Tat pathway. The ability of TorD to contribute significantly to the biosynthesis of an otherwise alien hydrogenase enzyme is quite surprising, and suggests that the proofreading mechanism employed by TorD can replicate almost exactly that of the native hydrogenase system. No known homologs of TorD are required for hydrogenase assembly. However, sequence analysis suggests TorD is homologous to DmsD.[31] DmsD has been shown to be essential for DMSO reductase assembly and to bind to the DMSO reductase twin-arginine signal peptide.[32] The original experiment of swapping the DmsA signal peptide for that of TorA would, therefore, also have removed an essential biosynthetic protein from the assembly process resulting in low DMSO reductase activity.[22]

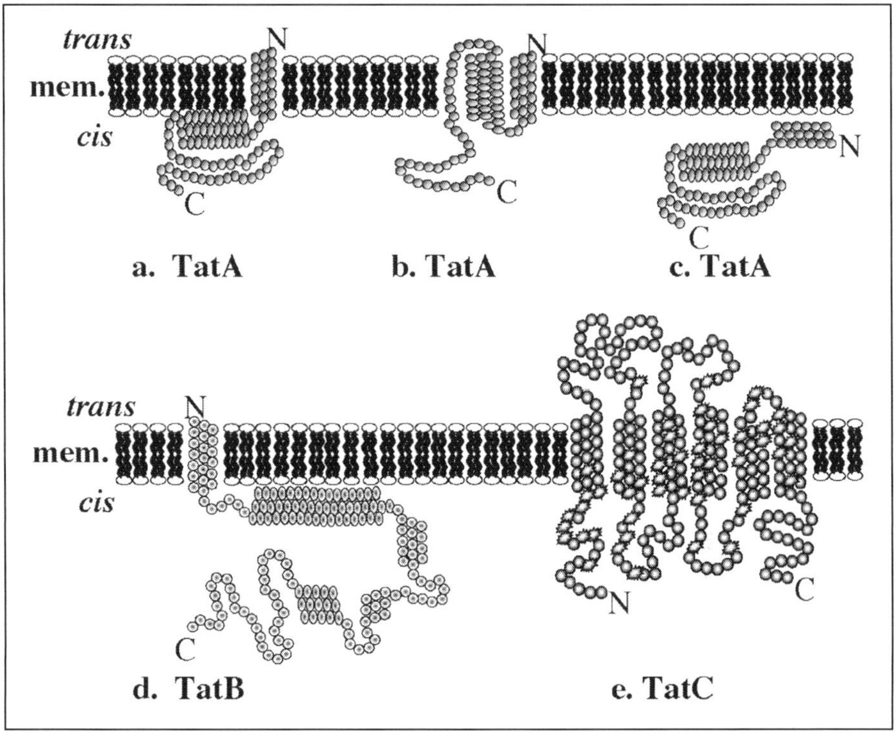

Figure 1. Predicted structures and topologies of Tat translocase components. a) the TatA protein as present in resting *E. coli* or thylakoid membranes;[2] b) alternative topology of the *E. coli* TatA protein as derived from PhoA fusions in living cells;[36] c) a coexisting pool of water-soluble TatA proteins has been reported in *B. subtilis*;[37] d) the TatB protein; and e) the polytopic integral membrane protein TatC recently confirmed to possess 6 transmembrane segments.[42]

At this juncture it is not clear whether such a chaperone-mediated proofreading mechanism is in operation for other Tat substrate proteins. However, it is interesting to note that the structural operons for many bacterial redox enzymes exported by the Tat pathway contain extra accessory genes that might encode such proteins.[31]

Tat Translocon Components

In general, three classes of membrane protein have been identified as components of the Tat translocase: TatA (also called Tha4 and Tha9 in the plant system), TatB (Hcf106 in plants), and TatC (Fig. 1). Genes encoding the three Tat components have been identified in a wide range of organisms[26] and mutations impair transport of all protein bearing twin-arginine signal peptides. Genes encoding Tat components are often duplicated[33] and *E. coli*, for example, produces two functional homologs of TatA (from the *tatA* and *tatE* genes[19]). TatA-class proteins are predicted to comprise a membrane-spanning α-helix at the N-terminus, immediately followed by a cytoplasmically-located amphipathic helix and then a C-terminal region of variable length (Fig. 1a). In most biological systems, TatA is bound to membranes,[34,35] probably via the N-terminus, although the overall topology[36] may be more complicated (Fig. 1b). *B. subtilis* contains three copies of TatA and studies of one of these ($TatA_d$[37]) suggests a portion of the protein may be soluble in the cell cytoplasm (Fig. 1c).

TatB-class proteins share a similar overall structure to TatA (Fig. 1). Studies in *E. coli*[38] and plants[39] indicate that TatB has a function in Tat transport that is distinct from that of TatA. The evolutionary link between TatA- and TatB-type proteins is so close, however, that some organisms, for example *B. subtilis*, contain no identifiable TatB homolog at all. In these cases, close inspection of the TatA protein sequence suggests the conserved features of both TatA and TatB have been combined in a single peptide indicating that the role of TatB may have been integrated with that of TatA.[40,41]

The TatC protein is highly hydrophobic and is predicted to have six transmembrane helices,[42] with the N- and C-termini located on the cytoplasmic face of the membrane (Fig. 1). TatC proteins show the highest level of amino acid conservation of all the Tat translocase components, with eight amino acids strictly conserved throughout the prokaryotic and eukaryotic TatC homologues.[43]

Phylogenetic analysis indicates that the bacterial *tatA-*, *tatB-*, *and tatC*-class genes very frequently show genetic linkage to each other but do not consistently cluster with any other types of genes. This suggests that TatABC probably form the only components of the Tat transport system. Indeed, overexpression of the *tatABC* genes alone results in a marked increase in protein flux through the *E. coli* Tat system both in vivo and in vitro.[44-47]

It is becoming clear that two separate large Tat complexes exist both in resting thylakoids and *E. coli* membranes—a TatA complex containing a low amount of TatB, and a complex containing equimolar amounts of the TatB and TatC proteins together with low and variable amounts of TatA.[35,44,48-52] Both complexes seem to be of the order of 400-700 kDa, implying multiple copies of each Tat component are present. Expression studies suggest that the ratio of TatA:TatB production is approximately 25:1, which is in good agreement with the levels of TatA and TatB found in membranes and in isolated TatAB complexes.[50,53] The membrane-bound TatAB complex from *E. coli*, which has been purified to apparent homogeneity, forms a large annulus when visualized by negative stain electron microscopy containing what may be a partially blocked cavity.[44] Purified TatBC complexes, on the other hand, are oval-shaped under negative stain electron microscopy and do not contain an internal cavity of note.[52]

Signal Peptide Recognition

The conserved twin-arginine motif within Tat signal peptides must be recognized by the Tat translocase for transport to proceed. The fact that Tat signal peptides are generally inactivated by mutating one or both motif arginines suggests the amino acid make-up of any binding pocket on the Tat translocase may also contain very highly conserved side-chains. It was originally speculated that the distinctive topologies of TatA- or TatB-class proteins might enable them to serve as 'receptors' for Tat substrates.[54] Indeed, this theory is borne-out to some extent by studies in *B. subtilis* that suggest a cytoplasmic pool of water-soluble TatA interacts with precursor proteins.[37] However, the primary sequence identity amongst these families of proteins is low and there are no conserved polar amino acids that might be expected to interact with the twin arginines of the signal peptide. Moreover, a pool of water-soluble TatA has not been reported during extensive studies of the *E. coli* system or the thylakoid pathway. In these systems, the TatC protein has emerged as the prime candidate for initial signal peptide recognition. Many of the conserved residues in TatC fall within a loop region between helices two and three which is exposed on the *cis* side of the membrane.[43] On this basis, it has been proposed that TatC may be the component of the transporter which recognizes the signal peptide. Further support for this contention came from a biophysical study of signal peptide binding to isolated TatBC complexes.[48] The most compellingly current evidence for the location of the signal binding site stems from recent in vitro experiments using isolated *E. coli* membranes containing elevated levels of Tat translocase components.[47] Using site-specific photocrosslinking technology, Alami and coworkers[47] were able to comprehensively demonstrate that side-chains within the twin-arginine motif of a Tat signal peptide were recognized directly by the TatC protein. Moreover, this motif recognition event was shown to be the first step in protein transport, the signal next being handed from TatC to TatB.[47] Thus, the TatBC complex is a membrane-located signal-binding module, a conclusion that is borne-out by complementary in vitro experiments with plant thylakoids.[51] As discussed below in more detail, interaction of the signal peptide with TatBC is probably via a 'loop' or 'hairpin' structure in which the extreme N-terminus of the signal peptide remains on the *cis* side of the membrane throughout the transport process.[44]

Protein Translocation

The unique role of the Tat system in post-translational transport of fully folded and assembled proteins means that three-dimensional structures of Tat substrates are important considerations when devising translocation theories. The smallest known Tat substrate is the 9 kDa high-potential iron-sulphur protein (HiPIP) of *Chromatium vinosum*, which has a maximum diameter of approximately 30Å[55] whilst the largest, *E. coli* formate dehydrogenase-N (FdnGH), has a diameter of around 70Å.[56] Thus, the Tat translocase must be capable of exporting substrate proteins that vary greatly in cross-sectional area and still maintain a tight seal in order to prevent leakage of protons and other ions across the cytoplasmic membrane during transport.

Negative-stain electron microscopy of the isolated *E. coli* TatAB complex results in the visualization of annular structures with a central cavity of 65Å in diameter.[44] The diameter of this cavity is of an appropriate order of magnitude to accommodate a folded Tat substrate protein and in vitro experiments suggest this TatA-rich complex could be the transport channel itself. Initially using isolated thlyakoids,[51] and more recently with *E. coli* membrane vesicles,[47] it was shown that the TatBC unit, once loaded with substrate and in the presence of a protonmotive force (PMF), associated with the TatA complex. Indeed, the *E. coli* in vitro experiments suggested that the twin-arginine signal peptide itself also moved close to TatA at this point, but only following initial recognition by TatBC.[47]

Thus, the experimental evidence points to a model in which the membrane-bound TatA complex is a protein-conducting channel, transiently interacting with the TatBC signal-binding module in a protonmotive force-dependent manner to facilitate substrate transport.[2,51] Clearly, transport of substrates by the Tat pathway is a dynamic event. Indeed, it has been reported that the C-terminal domain of TatA may, under some circumstances, be exposed at the periplasmic face of the membrane (Fig. 1b) and this raises the possibility of a dramatic 'flipping' of TatA topology, possibly during the transport event.[36]

Energy Transduction

It is likely that protein transport through the Tat channel is energized solely by the protonmotive force. While signal peptide recognition and binding by TatBC occurs efficiently even in the presence of protonophores, subsequent docking of the TatA complex is completely dependent upon PMF.[47,51] In *E. coli*, Tat transport rates can be boosted in vivo by overproduction of the phage shock protein—a membrane protein believed to be linked to PMF maintenance.[57] Moreover, Alder and Theg,[58] in in vitro studies of Tat transport in isolated thylakoid membranes, concluded that transport of a single OEC17 substrate protein was associated with retrograde transport of 80,000 protons. The number of protons involved seems extraordinarily large, and (at 3.66 protons required to synthesize 1 ATP) perhaps represents the equivalent of 22,000 ATPs lost to the chloroplast or plant cell. Could the bacterial Tat system operate to similar specifications? The energetics of the *E. coli* system have never been extensively tested, however assuming the volume of an *E. coli* cell is ~2 ×

10^{-19} litres (a cylinder of $r = 0.25$ μm and $h = 1$ μm) with an ATP concentration of ~5 mM, the total amount of ATP in an *E. coli* cell is ~1×10^{-21} moles or only ~600 individual molecules of ATP. Thus, despite the low level of predicted Tat transport sites in the membrane,[45,47] and even given an efficiently respiring cell, it may not be possible to perform bacterial Tat transport to such an energetically taxing regimen.

It should be noted, however, that Finazzi and coworkers[59] studied in vivo (as opposed to in vitro) Tat transport in chloroplasts from various sources but could find no dependence on the ΔpH component of PMF. The reasons for this observation are unclear. However a broad conclusion that could be made is that in vitro Tat transport assays may behave slightly differently to the situation in vivo. As Finazzi et al[59] point out, in vitro Tat transport is slow, incomplete, and for the bacterial system requires massive overproduction of Tat components. In contrast, in vivo Tat substrate precursors are processed so efficiently that they are difficult to detect, multiple Tat substrates are transported together without competition issues, the levels of active translocases, especially in Bacteria, is very low, and the energetic burden to the living cell is effortlessly balanced. Clearly, whilst we have witnessed recent great leaps in our understanding of protein transport, we have only scratched the surface with respect to learning the intricacies of the workings of the Tat translocase.

Post-Transport Events

Tat transport is very often not the final event in the biosynthesis of many proteins using this pathway. In most cases, the twin-arginine signal peptide must be removed, its job now complete. In addition, some Tat substrates contain transmembrane segments that must be integrated into the lipid bilayer and many more Tat targeted proteins are found to be homo- or hetero-oligomers upon purification. These final assembly steps probably occur after translocation and we will discuss each of them here in reverse order.

Long before the bacterial Tat translocase was described, Tat-dependent periplasmic redox enzymes have been isolated and characterized. The *E. coli* formate dehydrogenase-N, for example, is encoded by three genes—*fdnGHI*.[60] The FdnG catalytic subunit binds a molybdopterin cofactor and an Fe-S cluster and is synthesized with a twin-arginine signal peptide. FdnG forms a tight complex with the FdnH Fe-S which, in turn, is associated with the FdnI subunit—an integral membrane cytochrome *b*.[60] The crystal structure of formate dehydrogenase-N shows that the FdnGHI units oligomerize to form a $(FdnGHI)_3$ complex.[56] Numerous other examples of homo-oligomerization of Tat substrates exist (reviewed by Berks and colleagues[2]), including the homo-tetramer GFOR from *Z. mobilis*. The rationale for post-translocational rather than pretranslocational oligomerization is two-fold. Firstly, the Tat channel itself must ultimately be limited in terms of size of a potential Tat substrate. And secondly, oligomerization prior to transport would result in precursor proteins bristling with signal peptides which might complicate the transport process. How then does the cell prevent oligomerization before transport? The obvious physical difference between Tat precursors and mature proteins is the presence of the signal peptide and it has been suggested that the uncleaved signal itself might hinder oligomerization.[61] This theory arose from inspection of the $(FdnGHI)_3$ 'trimer-of-trimers' structure in which the FdnG extreme N-terminus was found to be completely buried at the trimer interface.[56] The presence of three 33 amino acid peptides in this region would undoubtedly disrupt any attempts at trimerization. Alternatively, specific signal-binding chaperones (e.g., TorD) or other cytoplasmic chaperones might be required for this process. This may partly explain why GFOR could not be heterologously exported by *E. coli* until an endogenous signal peptide was employed.[24]

As an interesting aside to this discussion, it is notable that some Tat-dependent proteins form stable periplasmic complexes with proteins targeted by alternative routes. The bacterial periplasmic nitrate reductase system, for example, comprises a Tat-targeted molybdoenzyme (NapA) and a non-Tat-targeted *c*-type cytochrome (NapB) that form a $NapA_1B_1$ complex. It is arguably astonishing to consider that the cell must coordinate translation, different export pathways, and cofactor biosynthesis and insertion pathways for two very different proteins and still somehow arrive at an active NapAB enzyme with perfect 1:1 stoichiometry.

The next post-transport event to consider is membrane protein integration. Recently, a subset of Tat substrate proteins have been revealed to be genuine integral membrane proteins containing at least one hydrophobic α-helix that completely spans the lipid bilayer.[12,61-63] Compelling evidence for the existence of Tat-dependent membrane proteins in Bacteria came again from *E. coli* formate dehydrogenase-N in which the Tat-dependent FdnH subunit was shown to contain a single transmembrane segment at its extreme C-terminus.[56] Subsequent biochemical work identified a further four such 'C-tail anchored' Tat substrates in *E. coli* and analogous proteins are widespread in Bacteria and plants.[63]

There are two possible mechanisms of membrane protein biosynthesis by the bacterial Tat translocase. Firstly, C-tail-anchors could behave as 'stop-transfer' domains. In the stop-transfer model, the hydrophobicity of the C-terminal helix would cause a stalling of the transport process as the helix passes through the Tat pore and the extreme C-terminus of the helix would, therefore, not enter the Tat channel. The now channel-located hydrophobic helix would then move laterally into the lipid bilayer. It is worth considering that such a stop-transfer model would probably require a channel-clearing mechanism.[61] It is possible such a 'stop-transfer' mechanism would require additional proteinaceous components in order to operate. However, none have so far been identified. It should be noted that under certain experimental conditions Tat-dependent membrane proteins have actually been recovered as soluble periplasmic intermediates.[64,65] While non-specific proteolytic degradation of the C-tail anchors could not be discounted in these experiments, it should perhaps be considered that periplasmic intermediates may play a role in the mechanism of Tat-dependent membrane protein biosynthesis. Integration could instead proceed by an alternative 'periplasmic reentry' mechanism and in this model no clearing of the Tat channel would be necessary since the entire enzyme, including the C-terminal helix, would be exported to the periplasm. The C-terminal tail-anchor would then associate with the inner membrane from the periplasmic side. The fact that Tat-dependent transmembrane segments are flanked by apparently random charged and hydrophilic side-chains suggests that any re-entry mechanism would also need to be protein-mediated.

As well as C-tail-anchored proteins, a second class of Tat-dependent integral membrane protein has been identified. The Rieske Fe-S protein of the cytochrome b_6f complex of the chloroplast thylakoid membrane is targeted in a Tat-dependent manner with the uncleaved Tat signal peptide doubling as a N-terminal signal-anchor.[12] Indeed, sequence and mutagenic analysis suggests that the analogous Rieske proteins found in the cytochrome bc_1 and b_6f complexes of many Bacteria may be assembled in the same way.[1,11,66]

Regardless of the integration mechanism employed, post-translational integration of membrane proteins is not without logistical problems. It may be necessary, for example, to prevent the hydrophobic transmembrane regions from engaging in non-specific aggregation processes prior to transport and final integration. In the case of Rieske-type N-terminal signal-anchors, the reduced hydrophobicity of the Tat signal anchor, when compared to standard transmembrane helices, would preclude any recognition by the signal recognition particle normally involved in membrane protein biosynthesis.[20] This should prevent mistargeting of most Tat substrates. However, we can find no significant differences in the overall hydrophobicity index between Tat-dependent C-tail-anchors and Sec-dependent transmembrane segments.[63] In this case, masking of the exposed hydrophobic helices could be another possible role for enzyme-specific accessory proteins.

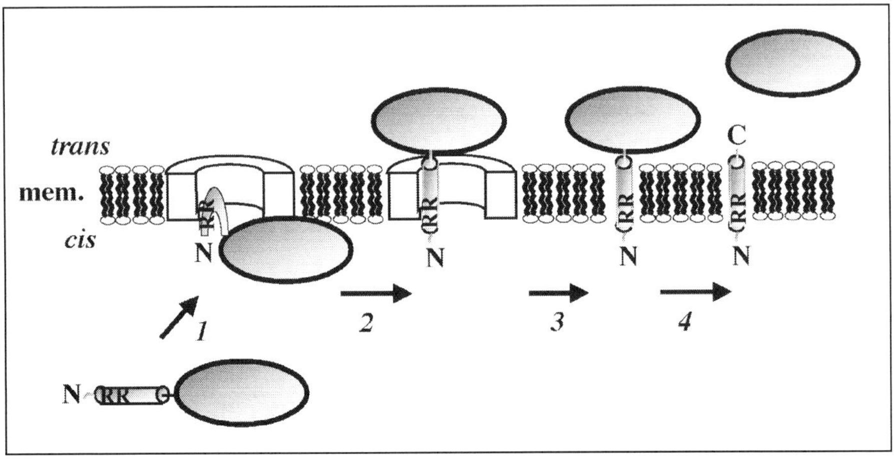

Figure 2. Targeting and transport by the Tat system. This model is based on in vitro work with *E. coli* membranes and chloroplast thylakoids.[47,51] *Step 1*, a Tat substrate is targeted to the membrane-bound Tat translocon by its signal peptide. The signal peptide initially interacts with TatC by forming a loop. *Step 2*, the passenger protein is transported. The event is energized by retrograde transport of protons. The signal peptide may remain inside the channel at this point with its N-terminus remaining on the *cis* side of the membrane. *Step 3*, lateral transfer of the signal into the lipid phase may occur inducing the signal peptide to adopt an α-helical conformation. For N-terminal signal-anchored proteins (e.g., Rieske) this is the end of the assembly process. *Step 4*, for signals bearing C-terminal AxA motifs, the signal peptide is finally cleaved off by a membrane-embedded protease (LepB in *E. coli*).

The existence of signal-anchored Tat substrates may also give some insight into the mechanism of the most common post-transport event—signal peptide cleavage. Most bacterial Tat signal peptides contain a conserved AxA amino acid motif (or an acceptable variation thereof) at the C-terminus.[1] This is the recognition site for LepB-type signal peptidases ('type I')[67] and use of a LepB-specific inhibitor in *E. coli* demonstrated unequivocally that this enzyme does indeed remove the SufI twin-arginine signal peptide.[45] *E. coli* LepB comprises a globular serine protease domain anchored to the periplasmic face of the cytoplasmic membrane by two N-terminal transmembrane helices. The active site pocket is on the "underside" of the protease domain and has been envisaged to 'skate' on the surface of the bilayer among the lipid headgroups. Signal peptide cleavage can, therefore, only occur, it seems, if the AxA motif is correctly oriented in the membrane. Cleavage would be impossible (not to mention undesirable) in the cytoplasm before export and would be difficult if the freshly transported precursor was free-floating in the periplasm. The most likely mechanism (which would also be in line with non-Tat soluble and integral membrane proteins processed by LepB) would involve initial insertion of the Tat signal peptide into the translocase in a 'loop' or 'hairpin' orientation[34] with N- and passenger-linked C-termini on the same side of the membrane (Fig. 2). Tat transport of the passenger domain would then result in a protein anchored by its N-terminal signal and probably still within the Tat channel (Fig. 2). Signal peptidases, including LepB, have never been copurified with any type of protein transport channel and, given the near pleotropic nature of signal cleavage in *E. coli*, it seems very unlikely that LepB would interact directly with the Tat tranlocase. In our model, the uncleaved signal would escape laterally through a 'gate' in the Tat channel and become a bona fide signal anchor in the lipid bilayer (Fig. 2). In this model, we envisage this hypothetical side gate would be identical to that used to integrate Tat-dependent transmembrane helices. Indeed, for signal-anchored Rieske Fe-S proteins this would be the final step in targeting. Lipid-embedded Tat signals would be expected to almost spontaneously form into α-helical structures[68] and this is certainly the case for the Rieske Fe-S protein, where the structure is known.[69] For all other Tat substrates, however, we finally have lipid-embedded Tat signals with AxA motifs correctly located within, or close to, the headgroups (Fig. 2). Cleavage would quickly follow. It is conceivable, therefore, that the apparently specialized mechanism of Tat-dependent transmembrane segment integration is arguably very intimately linked to the fundamental process of signal peptide cleavage.

Concluding Remarks

In the past 6 years we have witnessed the first wave of research into the Tat transport system. The basic components have been defined and vital tools, such as in vitro assays and purification protocols, have been developed. Now the race is well-and-truly underway for the first structural information on this transport system which, once available, will spark the eagerly awaited new wave of Tat research where we will begin at last to understand the mechanism of this most remarkable molecular machine.

References

1. Berks BC. A common export pathway for proteins binding complex redox cofactors? Mol Microbiol 1996; 22:393-404.
2. Berks BC, Palmer T, Sargent F. The Tat protein translocation pathway and its role in microbial physiology. Adv Microb Physiol 2003; 47:187-254.
3. Santini CL, Ize B, Chanal A et al. A novel Sec-independent periplasmic protein translocation pathway in Escherichia coli. EMBO J 1998; 17:101-112.
4. Rodrigue A, Chanal A, Beck K et al. Cotranslocation of a periplasmic enzyme complex by a hitchhiker mechanism through the bacterial Tat pathway. J Biol Chem 1999; 274:13223-13228.
5. Buchanan G, Kuper J, Mendel RR et al. Characterisation of the mob locus of Rhodobacter sphaeroides WS8: Moba is the only gene required for molybdopterin guanine dinucleotide synthesis. Arch Microbiol 2001; 176:62-68.
6. Stanley NR, Palmer T, Berks BC. The twin arginine consensus motif of Tat signal peptides is involved in Sec-independent protein targeting in Escherichia coli. J Biol Chem 2000; 275:11591-11596.
7. Ize B, Stanley NR, Buchanan G et al. Role of the Escherichia coli Tat pathway in outer membrane integrity. Mol Microbiol 2003; 48:1183-1193.
8. Bernhardt TG, de Boer PA. The Escherichia coli amidase AmiC is a periplasmic septal ring component exported via the twin-arginine transport pathway. Mol Microbiol 2003; 48:1171-1182.
9. Faury D, Saidane S, Li H et al. Secretion of active xylanase C from Streptomyces lividans is exclusively mediated by the Tat protein export system. Biochim Biophys Acta 2004; 1699:155-162.
10. DeLisa MP, Tullman D, Georgiou G. Folding quality control in the export of proteins by the bacterial twin-arginine translocation pathway. Proc Natl Acad Sci USA 2003; 100:6115-6120.
11. Hinsley AP, Stanley NR, Palmer T et al. A naturally occurring bacterial Tat signal peptide lacking one of the 'invariant' arginine residues of the consensus targeting motif. FEBS Lett 2001; 497:45-49.
12. Molik S, Karnauchov I, Weidlich C et al. The Rieske Fe/S protein of the cytochrome b6/f complex in chloroplasts: Missing link in the evolution of protein transport pathways in chloroplasts? J Biol Chem 2001; 276:42761-42766.
13. Gross R, Simon J, Kröger A. The role of the twin-arginine motif in the signal peptide encoded by the hydA gene of the hydrogenase from Wolinella succinogenes. Arch Microbiol 1999; 172:227-232.
14. Halbig D, Wiegert T, Blaudeck N et al. The efficient export of NADP-containing glucose-fructose oxidoreductase to the periplasm of Zymomonas mobilis depends both on an intact twin-arginine motif in the signal peptide and on the generation of a structural export signal induced by cofactor binding. Eur J Biochem 1999; 263:543-551.

15. Santini CL, Bernadac A, Zhang M et al. Translocation of jellyfish green fluorescent protein via the TAT system of Escherichia coli and change of its periplasmic localization in response to osmotic up-shock. J Biol Chem 2001; 276:8159-8164.
16. Thomas JD, Daniel RA, Errington J et al. Export of active green fluorescent protein to the periplasm by the twin-arginine translocase (Tat) pathway in Escherichia coli. Mol Microbiol 2001; 39:47-53.
17. DeLisa MP, Samuelson P, Palmer T et al. Genetic analysis of the twin-arginine translocator secretion pathway in bacteria. J Biol Chem 2002; 277:29825-29831.
18. Ize B, Gerard F, Zhang M et al. In vivo dissection of the Tat translocation pathway in Escherichia coli. J Mol Biol 2001; 317:327-335.
19. Sargent F, Bogsch EG, Stanley NR et al. Overlapping functions of components of a bacterial Sec-independent protein export pathway. EMBO J 1998; 17:3640-3650.
20. Cristóbal S, de Gier JW, Nielsen H et al. Competition between Sec- and Tat-dependent protein translocation in Escherichia coli. EMBO J 1999; 18:2982-2990.
21. Blaudeck N, Kreutzenbeck P, Freudl R et al. Genetic analysis of pathway specificity during post-translational protein translocation across the Escherichia coli plasma membrane. J Bacteriol 2003; 185:2811-2819.
22. Sambasivarao D, Turner RJ, Simala-Grant J et al. Multiple roles for the twin arginine leader sequence of dimethyl sulfoxide reductase of Escherichia coli. J Biol Chem 2000; 275:22526-22531.
23. Jack RL, Buchanan G, Dubini A et al. Coordinating assembly and export of complex bacterial proteins. EMBO J 2004; 23:3962-3972.
24. Blaudeck N, Sprenger GA, Freudl R et al. Specificity of signal peptide recognition in Tat-dependent bacterial protein translocation. J Bacteriol 2001; 183:604-610.
25. Buchanan G, Sargent F, Berks BC et al. A genetic screen for suppressors of Escherichia coli Tat signal peptide mutations establishes a critical role for the second arginine within the twin-arginine motif. Arch Microbiol 2001; 177:107-112.
26. Berks BC, Sargent F, Palmer T. The Tat protein export pathway. Mol Microbiol 2000; 35:260-274.
27. Henry R, Carrigan M, McCaffrey M et al. Targeting determinants and proposed evolutionary basis for the Sec and the Delta pH protein transport systems in chloroplast thylakoid membranes. J Cell Biol 1997; 136:823-832.
28. Niviere V, Wong SL, Voordouw G. Site-directed mutagenesis of the hydrogenase signal peptide consensus box prevents export of a beta-lactamase fusion protein. J Gen Microbiol 1992; 138:2173-2183.
29. Ilbert M, Mejean V, Iobbi-Nivol C. Functional and structural analysis of members of the TorD family, a large chaperone family dedicated to molybdoproteins. Microbiology 2004; 150:935-943.
30. Ilbert M, Mejean V, Giudici-Orticoni MT et al. Involvement of a mate chaperone (TorD) in the maturation pathway of molybdoenzyme TorA. J Biol Chem 2003; 278:28787-28792.
31. Turner RJ, Papish AL, Sargent F. Sequence analysis of bacterial redox enzyme maturation proteins (REMPs). Can J Microbiol 2004; 50:225-238.
32. Oresnik IJ, Ladner CL, Turner RJ. Identification of a twin-arginine leader binding protein. Mol Microbiol 2001; 40:323-331.
33. Yen MR, Tseng YH, Nguyen EH et al. Sequence and phylogenetic analyses of the twin-arginine targeting (Tat) protein export system. Arch Microbiol 2001; 177:441-450.
34. Fincher V, McCaffery M, Cline K. Evidence for a loop mechanism of protein transport by the thylakoid delta pH pathway. FEBS Lett 1998; 423:66-70.
35. Porcelli I, de Leeuw E, Wallis R et al. Characterization and membrane assembly of the TatA component of the Escherichia coli twin-arginine protein transport system. Biochemistry 2002; 41:13690-13697.
36. Gouffi K, Gerard F, Santini CL et al. Dual topology of the Escherichia coli TatA protein. J Biol Chem 2004; 279:11608-11615.
37. Pop OI, Westermann M, Volkmer-Engert et al. Sequence-specific binding of prePhoD to soluble TatAd indicates protein-mediated targeting of the Tat export in Bacillus subtilis. J Biol Chem 2003; 278:38428-38436.
38. Sargent F, Stanley NR, Berks BC et al. Sec-independent protein translocation in Escherichia coli: A distinct and pivotal role for the TatB protein. J Biol Chem 1999; 274:36073-36083.

39. Walker MB, Roy LM, Coleman E et al. The maize tha4 gene functions in sec-independent protein transport in chloroplasts and is related to hcf106, tatA, and tatB. J Cell Biol 1999; 147:267-276.
40. Hicks MG, de Leeuw E, Porcelli I et al. The Escherichia coli twin-arginine translocase: Conserved residues of TatA and TatB family components involved in protein transport. FEBS Lett 2003; 539:61-67.
41. Lee PA, Buchanan G, Stanley NR et al. Truncation analysis of TatA and TatB defines the minimal functional units required for protein translocation. J Bacteriol 2002; 184:5871-5879.
42. Behrendt J, Standar K, Lindenstrauss U et al. Topological studies on the twin-arginine translocase component TatC. FEMS Microbiol Lett 2004; 234:303-308.
43. Buchanan G, Leeuw E, Stanley NR et al. Functional complexity of the twin-arginine translocase TatC component revealed by site-directed mutagenesis. Mol Microbiol 2002; 43:1457-14570.
44. Sargent F, Gohlke U, de Leeuw E et al. Purified components of the Tat protein transport system of Escherichia coli form a double-layered ring structure. Eur J Biochem 2001; 268:3361-3367.
45. Yahr TL, Wickner WT. Functional reconstitution of bacterial Tat translocation in vitro. EMBO J 2001; 20:2472-2479.
46. Alami M, Trescher D, Wu LF et al. Separate analysis of twin-arginine translocation (Tat)-specific membrane binding and translocation in Escherichia coli. J Biol Chem 2002; 277:20499-20503.
47. Alami M, Lüke I, Deitermann S et al. Differential interactions between a twin-arginine signal peptide and its translocase in Escherichia coli. Mol Cell 2003; 12:937-946.
48. de Leeuw E, Granjon T, Porcelli I et al. Oligomeric properties and signal peptide binding by Escherichia coli Tat protein transport complexes. J Mol Biol 2002; 322:1135-146.
49. Bolhuis A, Bogsch EG, Robinson C. Subunit interactions in the twin-arginine translocase complex of Escherichia coli. FEBS Lett 2000; 472:88-92.
50. Bolhuis A, Mathers JE, Thomas JD et al. TatB and TatC form a functional and structural unit of the twin-arginine translocase from Escherichia coli. J Biol Chem 2001; 276:20213-20219.
51. Cline K, Mori H. Thylakoid DeltapH-dependent precursor proteins bind to a cpTatC-Hcf106 complex before Tha4-dependent transport. J Cell Biol 2001; 154:719-729.
52. Oates J, Mathers J, Mangels D et al. Consensus structural features of purified bacterial TatABC complexes. J Mol Biol 2003; 330:277-286.
53. Jack RL, Sargent F, Berks BC et al. Constitutive expression of the Escherichia coli tat genes indicates an important role for the twin-arginine translocase during aerobic and anaerobic growth. J Bacteriol 2001; 183:1801-1804.
54. Chanal A, Santini CL, Wu LF. Potential receptor function of three homologous components, TatA, TatB and TatE, of the twin-arginine signal sequence-dependent metalloenzyme translocation pathway in Escherichia coli. Mol Microbiol 1998; 30:674-676.
55. Carter Jr CW, Kraut J, Freer ST et al. Two-angstrom crystal structure of oxidized Chromatium High Potential Iron Protein. J Biol Chem 1974; 249:4214.
56. Jormakka M, Tornroth S, Byrne B et al. Molecular basis of proton motive force generation: Structure of formate dehydrogenase-N. Science 2002; 295:1863-1868.
57. DeLisa MP, Lee P, Palmer T et al. Phage shock protein PspA of Escherichia coli relieves saturation of protein export via the Tat pathway. J Bacteriol 2004; 186:366-373.
58. Alder NN, Theg SM. Energetics of protein transport across biological membranes: A study of the thylakoid DeltapH-dependent/cpTat pathway. Cell 2003; 112:231-242.
59. Finazzi G, Chasen C, Wollman FA et al. Thylakoid targeting of Tat passenger proteins shows no delta pH dependence in vivo. EMBO J 2003; 22:807-815.
60. Berg BL, Li J, Heider J et al. Nitrate-inducible formate dehydrogenase in Escherichia coli K-12. I. Nucleotide sequence of the fdnGHI operon and evidence that opal (UGA) encodes selenocysteine. J Biol Chem 1991; 266:22380-22385.
61. Sargent F, Berks BC, Palmer T. Assembly of membrane-bound respiratory complexes by the Tat protein-transport system. Arch Microbiol 2002; 178:77-84.
62. Summer EJ, Mori H, Settles AM et al. The thylakoid delta pH-dependent pathway machinery facilitates RR-independent N-tail protein integration. J Biol Chem 2000; 275:23483-23490.
63. Hatzixanthis K, Palmer T, Sargent F. A subset of bacterial inner membrane proteins integrated by the twin-arginine translocase. Mol Microbiol 2003; 49:1377-1390.

64. Bernhard M, Benelli B, Hochkoeppler A et al. Functional and structural role of the cytochrome b subunit of the membrane-bound hydrogenase complex of Alcaligenes eutrophus H16. Eur J Biochem 1997; 248:179-186.
65. Stanley NR, Sargent F, Buchanan G et al. Behaviour of topological marker proteins targeted to the Tat protein transport pathway. Mol Microbiol 2002; 43:1005-1021.
66. Meloni S, Rey L, Sidler S et al. The twin-arginine translocation (Tat) system is essential for Rhizobium-legume symbiosis. Mol Microbiol 2003; 48:1195-1207.
67. Paetzel M, Dalbey RE, Strynadka NC. Crystal structure of a bacterial signal peptidase in complex with a beta-lactam inhibitor. Nature 1998; 396:186-190.
68. San Miguel M, Marrington R, Rodger PM et al. An Escherichia coli twin-arginine signal peptide switches between helical and unstructured conformations depending on the hydrophobicity of the environment. Eur J Biochem 2003; 270:3345-3352.
69. Kurisu G, Zhang H, Smith JL et al. Structure of the cytochrome b_6f complex of oxygenic photosynthesis: Tuning the cavity. Science 2003; 302:1009-1014.

CHAPTER 7

Retro-Translocation of Proteins Across the Endoplasmic Reticulum Membrane

J. Michael Lord* and Lynne M. Roberts

Abstract

Many proteins synthesised in the cytosol are translocated across or inserted into the endoplasmic reticulum (ER) membrane. These proteins include not only those resident in the ER itself, but others destined for post-ER destinations such as the Golgi complex, lysosomes or secretion into the extracellular environment. Proteins that fail to fold or assemble correctly are detected by the quality control system of the ER and are disposed of by a process known as ER-associated degradation. Degradation does not occur in the ER itself. Rather the aberrant proteins are exported from the ER for degradation by the ubiquitin/proteasome pathway in the cytosol. This involves the retro-translocation of these proteins across the ER membrane. In this chapter we discuss our current understanding of the process of retro-translocation.

A significant proportion of the proteins synthesised by ribosomes in the cytosol are translocated across or inserted into the membrane of the endoplasmic reticulum (ER).[1] These proteins include not only the luminal and membrane proteins resident in the ER itself, but also others that are transported beyond the ER, such as secretory, plasma membrane, lysosomal and Golgi proteins.[1] All use the same passive conduits to cross or insert into the ER membrane: protein-conducting channels termed translocons.[2] Conserved heterotrimeric membrane proteins known as Sec61p or the Sec61 complex in eukaryotes form oligomers that constitute the core of these translocons. Delivery of polypeptides into the ER occurs during or soon after their synthesis[1,3] and, to ensure that translocated proteins become biologically functional, it is essential that they fold correctly and, if part of oligomeric complexes, assemble correctly. These processes are facilitated by a range of soluble and membrane-associated molecular chaperones.[4] Proteins fold and assemble into multimeric structures in step-wise fashion, assisted by sequential interactions with specific chaperones that may be ubiquitous or cell-type specific. The ER lumen also contains enzymes that can modify particular nascent proteins by catalyzing disulfide bond formation, glycosylation, and/or the addition of lipid anchors. Throughout the folding and maturation process, proteins are monitored by an elaborate and stringent ER quality control system that employs various folding sensors and molecular chaperones to ensure 'native' conformations are reached.[5,6] On the whole, proteins

*Corresponding Author: J. Michael Lord—Department of Biological Sciences, University of Warwick, Coventry CV4 7AL, U.K. Email: mike.lord@warwick.ac.uk

Protein Movement Across Membranes, edited by Jerry Eichler. ©2005 Eurekah.com and Springer Science+Business Media.

destined for post-ER locations are only allowed to exit the ER in transport vesicles when they pass this test of structural integrity.

Mutations and missing subunits can lead to aberrantly folded proteins. Although some of these may be trafficked to lysosomes for degradation,[7] most are not dispatched to their intended destinations, nor do they usually accumulate indefinitely. Initially it was believed that such proteins were degraded by proteases resident in the ER,[8] but it is now accepted that misfolded proteins are exported from the ER and degraded by the ubiquitin/proteasome system in the cytosol.[9] Terminally misfolded proteins must therefore undergo retro-translocation across the ER membrane before they can be tagged by ubiquitin and degraded. This aspect of ER quality control is known as ER-associated degradation (ERAD).[9,10] It is a process that must be highly selective in order to distinguish between folding intermediates and those proteins that are terminally misfolded.

ERAD leading to complete degradation is not an obligatory fate of aberrant ER proteins however. Some may accumulate or aggregate, a situation found in a variety of human diseases.[11,12] Inefficient disposal in the cystosol can lead to the formation of accumulations of ubiquitinated proteins called aggresomes,[13] while inefficient retro-translocation or an overwhelming demand for it can lead to ER accumulations of heavily aggregated mutant proteins that form dilated ER cisternae known as Russell bodies.[13]

Ever changing demands placed on the ER have led to the evolution of regulatory mechanisms to constantly monitor and adjust the levels of available molecular chaperones. In this regard, accumulation of misfolded proteins in the ER will trigger a cellular stress-induced signalling pathway known as the unfolded protein response (UPR)[14] that up-regulates expression of genes encoding ER proteins.[15] Indeed, a regulatory link between ERAD and the UPR has been demonstrated.[16] The mammalian transducers of the UPR, i.e., PERK, ATF6 and Ire1,[17] are kept inactive by interactions with chaperones such as BiP. PERK is a transmembrane kinase that phosphorylates eIF2 to attenuate global translation and relieve the load of continued protein synthesis, and to arrest the cell cycle. ATF6 is a transmembrane-activating transcription factor that, upon ER stress, is transported to the Golgi where it becomes cleaved to release a cytosolic fragment. This is turn is transported to the nucleus to activate transcription. Ire1 is a transmembrane protein in which both a kinase and ribonuclease activity are contained in its cytosolic domain. Once activated by ER stress, Ire1 catalyses splicing of XBP1 mRNA that encodes a transcription factor. The mammalian stress element is present in the promoters of UPR-targeted genes, many of which encode proteins that mediate ERAD. BiP is the master regulator of the activation of PERK, ATF6 and Ire1 through its interaction with their luminal domains. Under normal conditions, BiP maintains the UPR transducers in inactive form. Upon ER stress, BiP binds to unfolded proteins and is thereby titrated from Ire1 and PERK to allow their homodimerization and activation. Release of BiP from ATF6 allows its transit to the Golgi where it becomes proteolytically activated. This BiP-regulated activation therefore allows the protein folding capacity of the ER to be increased once chaperone insufficiency is sensed.

The first abnormal proteins to be studied in terms of their cellular fate were orphan subunits of the T cell receptor.[18-20] The mature antigen receptor is an oligomeric complex of transmembrane proteins on the surface of T cells. Individual subunits of the T cell receptor are synthesised at different rates so the amount of mature receptor is defined by the least abundant subunit. Partial complexes or free subunits did not appear on the cell surface but were found to be degraded in a non-lysosomal, pre-Golgi compartment, as noted above, initially thought to be the ER itself.[8,21] However, the existence of a vigorous proteolytic system in an environment dedicated to the correct folding of proteins was difficult to reconcile. Cytosolic proteasomes were known to be a major site of protein degradation in the cell, with proteolytic substrates being marked for degradation by the covalent attachment of ubiquitin, catalyzed by ubiquitin-conjugating enzymes.[22] The first indication that ER proteins might be degraded by the ubiquitin/proteasome system came from genetic studies using yeast. In 1993, Sommer and Jentsch reported the integral membrane ubiquitin-conjugating enzyme Ubc6p was present at the cytoplasmic surface of the ER membrane. *UBC6* loss-of-function mutants suppressed a protein translocation defect caused by a mutation in *SEC61.* They concluded that Ubc6p might mediate the cytosolic degradation of ER membrane proteins by the proteasome, and that this was the fate of the mutant Sec61p.[23]

Support for the contention that abnormal ER proteins can be degraded by the proteasome has since come from many studies of aberrant membrane and soluble proteins (reviewed in ref. 24). In some instances, however, normal proteins are retained and degraded by quality control as exemplified by the fate of wild-type and mutant cystic fibrosis transmembrane conductance regulator (CFTR). For reasons that are unclear, only 25% of native CFTR inserted into the ER membrane is trafficked to its site of function at the plasma membrane. The remaining 75% of the native protein is degraded in a pre-Golgi compartment. This compares with 100% degradation of the ΔF508 mutant CFTR that is found in most cystic fibrosis patients. The degradation of both wild-type and mutant CFTR was shown to occur by the ubiquitin/proteasome pathway[25,26] In certain circumstances, ERAD is responsible for the physiologically regulated degradation of native proteins. The classical example of this is in the regulation of cholesterol synthesis. Hydroxymethylglutaryl-coenzyme A reductase (HMGR) is a rate-limiting enzyme of the mevalonate pathway by which sterols such as cholesterol and other isoprenoids are synthesized. When high cellular cholesterol levels are evident, HMGR is degraded by the ubiquitin-proteasome pathway to prevent further accumulation of this sterol.[27,28]

Certain viruses subvert ERAD as a strategy for immune evasion. For example, expression of the human cytomegalovirus (HCMV) genes US2 and US11 leads to the degradation of the host cell major histocompatibility complex (MHC) class I heavy chains. The viral gene products cause the dislocation of MHC molecules from the ER membrane and their export to the cytosol where they are ubiquitinated and degraded.[29,30] When the cells were treated with proteasome inhibitors, the dislocated MHC class I heavy chains accumulated in the cytosol as soluble, ubiquitinated, deglycosylated proteins.[29]

Interestingly, while it was initially assumed that misfolded ER proteins would never exit this organelle, it has recently been shown that some misfolded/unassembled proteins can be cycled to the Golgi for subsequent retrieval prior to retro-translocation and degradation.[31-35] Initial studies in yeast suggested that ERAD substrates requiring ER-to-Golgi transport for degradation were typically soluble proteins, whereas membrane substrates were retained in the ER before retro-translocation and degradation.[31,36] This model may be an oversimplification based on study of a limited number of substrates. It has recently been subjected to more rigorous testing, with the finding that cellular decision to statically retain the protein in the ER or subject it to ER-to-Golgi cycling prior to degradation does not depend on association with the membrane but rather on the site of the lesion causing misfolding.[37] It has been shown that distinct topological domains of the substrate polypeptides were subjected to two sequential checkpoints. The first examines the cytosolic domain of ER membrane proteins. If a lesion is found in this domain, the protein is retained in the ER and rapidly degraded regardless of the state of other domains. Proteins passing this first check are then subjected to a second that examines the ER luminal domain. Proteins detected by this second checkpoint are sorted from correctly folded proteins and undergo ER-to-Golgi transport prior to degradation. While the first checkpoint is exclusive to membrane proteins, the second monitors both membrane and soluble proteins.[37]

How is retro-translocation across the ER membrane to the cytosol achieved? Clearly the first step is to recognise that ERAD substrates are misfolded/unassembled and to target them to the retro-translocation machinery. Misfolded proteins tend to expose hydrophobic regions that would be buried in the native conformation. During normal protein folding in the ER, resident molecular chaperones bind to transiently exposed hydrophobic regions to prevent unproductive interactions between these regions.[38] It is therefore not surprising that ER molecular chaperones also bind ERAD substrates.[10,39-41] It has been suggested that the role of ER chaperone binding is to prevent extensive aggregation of ERAD substrates, thereby maintaining them in a retro-translocation-competent state.[42] If chaperones transiently bind to both wild type proteins in the act of folding and mutant proteins incapable of reaching their native conformations, a clear distinction must be made between the two. Precisely how this distinction is made in all cases is unclear at present, but it may be determined by the time taken to fold correctly. That is, chaperones may associate with proteins long enough to allow the correct folding, beyond which any proteins remaining misfolded are delivered to the retro-translocation machinery.

In mammalian cells, the calnexin cycle, that is responsible for effecting the folding of glycoproteins with N-linked oligosaccharides in the ER, utilizes a timing mechanism based on the structure of the glycan moiety (Fig. 1).[6,24] Core glycosylation in the ER results in the addition of an oligosaccharide consisting of two *N*-acetylglucosamine residues, nine mannose residues and three glucose residues ($((GlcNac)_2(Man)_9(Glu)_3)$ to nascent glycoproteins. Glucosidases I and II in the ER lumen rapidly remove two of the three glucose residues from newly synthesised but unfolded glycoproteins. This allows the monoglucosylated glycan to interact with the ER membrane-bound lectin calnexin. Glucosidase II removes the remaining glucose and the glycoprotein is rapidly released from calnexin. If the released glycoprotein is correctly folded, it can leave the ER via the secretory pathway. If it remains imperfectly folded, it is recognised by an UDP-glucose:glycoprotein glucosyltransferase (UGGT), a folding sensor, and is once more monoglucosylated.[43] This allows iterative rounds of binding to calnexin to provide further opportunities for folding. Terminally misfolded glycoproteins are marked for degradation by the action of mannosidase I which opportunistically removes one or more mannose residues to generate a $(GlcNac)_2(Man)_{8/7}$ form of the glycan. Mannose trimming lowers the activity of UGGT,[44] ultimately reducing further interactions with calnexin. Indeed, studies support the idea that the mannose-trimmed protein can be extracted from calnexin by an α-mannosidase I-like lectin known as EDEM[45,46] (ER degradation enhancing α-mannose-like protein) and diverted into the ERAD pathway. Mannosidase generation of the $(GlcNac)_2(Man)_8$ glycan form is relatively slow, so only proteins that fail to fold correctly after a period of time are thought to be targeted for degradation in this way. However, not all glycoproteins fold with the same kinetics, so this timing may be flexible or it may be the case that multiple timers exist. Whether EDEM directly hands calnexin-selected ERAD substrates to the retro-translocation apparatus isn't known. Nor it is yet clear what timing mechanism operates for misfolded non-glycosylated proteins, and for the ER substrates in yeast where calnexin and UGGT are absent,[24] even though the ERAD of glycoproteins is mannosidase 1-dependent.[47] Indeed, the calnexin/calreticulin cycle cannot be the only pathway for the ERAD of glycoproteins since mannosidase-dependent degradation can occur in some instances without calnexin or UGGT.[48]

Other chaperones involved in targeting ERAD substrates to the degradation pathway include BiP, an Hsp70 family member that associates with unassembled immunoglobulin chains before their degradation.[49,50] In yeast, mutations in the BiP homologue Kar2 block the degradation of several ERAD substrates.[51,52] Protein disulfide isomerase (PDI), an ER luminal protein with multiple functions,[53] is also involved in ERAD. Export from the ER of a cysteine-free misfolded secretory protein for cytosolic degradation in yeast required its interaction with PDI,[54] although other proteins apparently require reduction of disulfide bonds to become substrates for ERAD.[55] Eps1p, a novel yeast membrane protein belonging to the PDI family, was required for ERAD of a mutant plasma membrane protein.[56,57]

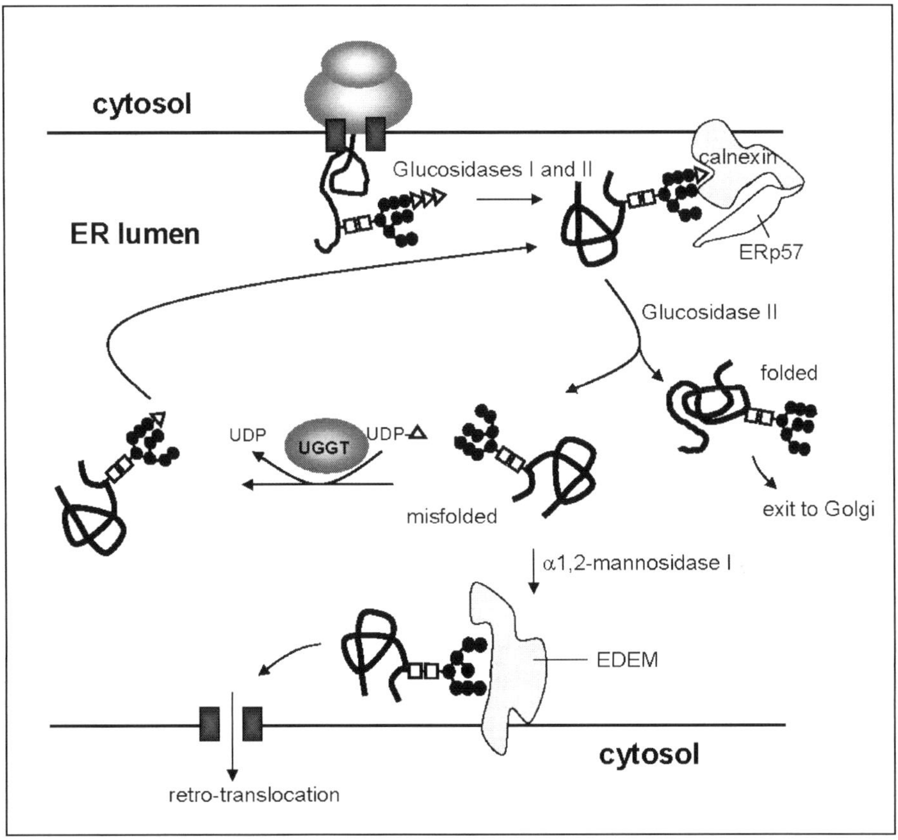

Figure 1. The calnexin-calreticulin quality control cycle in higher eukaryotes. Following the action of ER glucosidases, monoglucosylated glycoproteins with structural defects will bind the monoglucose-specific lectins, calnexin (shown) or soluble calreticulin that in turn interact with the thiol-disulphide oxidoreductase ERp57 (through which disulphide bonds may form). This binding of newly imported proteins with calnexin/calreticulin slows down their rate of folding but increases its efficiency. Release from calnexin occurs when glucosidase II cleaves the remaining glucose at which point correctly folded proteins will be permitted to exit the ER for the Golgi. Proteins that have not yet reached their native conformations will be detected by the folding sensor UDP-glucose:glycoprotein glucosyltransferase (UGGT) that can reglucosylate the protein. This allows a further interaction with calnexin/calreticulin in a cycle that may be repeated many times. If the protein fails to reach its native conformation by the time the slow acting ER α,2-mannosidase I removes a single mannose from the oligosaccharide, the $Glc_1Man_8GlcNAc_2$ glycan structure is recognised by ER degradation-enhancing 1,2-mannosidase-like protein (EDEM). This may target the glycoprotein for retro-translocation.

After a protein is identified as an ERAD substrate, it is exported from the ER. Experiments showing mammalian ER export of dislocated MHC class I heavy chains mediated by the product of the HCMV US2 gene,[30] and studies with specific yeast mutants,[52,58] first suggested that export in both systems involved the Sec61 translocon. This appeared to involve a reversal of the process by which nascent secretory proteins are delivered into the ER lumen. Indeed, mutant forms of Sec61p itself are degraded by the ubiquitin-proteasome pathway, presumably after retro-translocation via functional Sec61 complexes.[59] In order to fit into the channel, it seems likely that most substrates would be at least partially unfolded, for, even at its widest

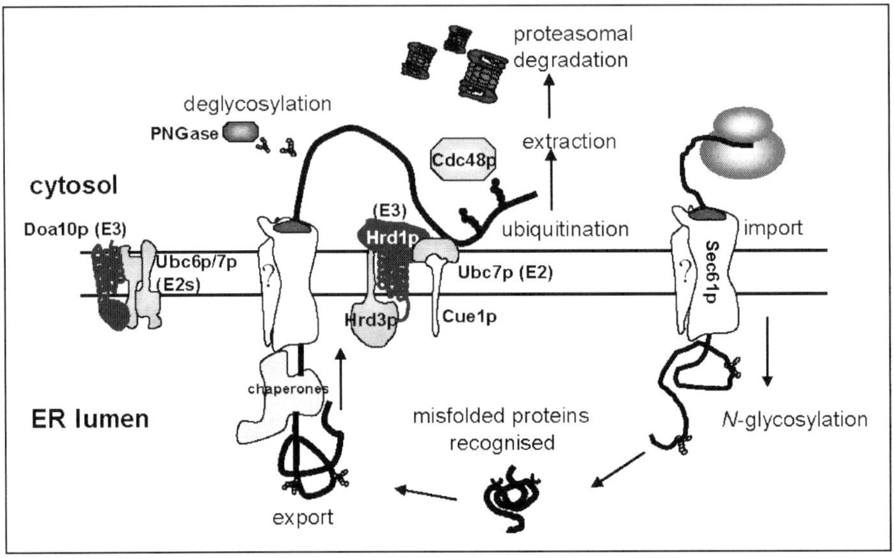

Figure 2. Schematic of the ER-associated protein degradation (ERAD) pathway for soluble misfolded proteins in *Saccharomyces cerevisiae*. Newly made proteins are delivered through Sec61p translocons into the endoplasmic reticulum lumen. They may be cycled through the Golgi (not shown) before being targeted to ER translocons by ER quality control proteins for retro-translocation to the cytosol. ER proteins including the chaperone Kar2p, mannose-binding lectins and several members of the protein disulphide isomerase family have been implicated in the turnover of soluble ERAD substrates. Membrane proteins (e.g., Hrd3p and Der1p (not shown)) are believed to target misfolded proteins to the translocon or its accessory protein (denoted by ?) The ubiquitin-conjugating enzymes (E2s) (Ubc1p, Ubc6p and Ubc7p), and membrane bound ubiquitin ligases (E3s) (Hrd1p and Doa10p) are responsible for the polyubiquitination of internal lysyl residues of yeast ERAD substrates. Cue1p recruits soluble Ubc7p to the ER membrane. Ubiquitination of ERAD substrates on the cytosolic surface of the membrane appears to be crucial for the retro-translocation process, most likely for the recruitment of other complexes such as proteasomes or the Cdc48p complex (Cdc48p/No14p/Ufd1p) that facilitate extraction from the membrane. The polyubuiquitin chains signal recognition, unfolding and degradation by proteasomes. PNGase is a cytosolic peptide:*N*-glycanase that removes glycans from glycoproteins at some point prior to their degradation. For clarity, not all proteins are shown in the scheme.

estimate (40-60Å diameter),[60] large glycoproteins would not be accommodated without some degree of unfolding. Precisely how ERAD substrates are brought to the Sec61 channel and threaded through it remains unclear. It also remains possible that translocons in distinct ER subcompartments are responsible for retro-translocation.

Most substrates that are retro-translocated become polyubiquitinated during this process. Many of the components of this system have been identified in yeast and show that ubiquitin conjugating enzymes Ubc1, membrane-bound Ubc6, and Ubc7 are involved (Fig. 2). Mutations in these proteins result in the stabilization of particular substrates (reviewed in ref. 61). These findings have given rise to the idea that polyubiquitin may provide the driving force for moving the proteins across the membrane by a ratcheting mechanism. Alternatively, extraction of ERAD substrates may be achieved by a different molecule. Several studies have implicated a protein complex at the ER membrane containing the yeast AAA-ATPase Cdc48 (p97 in mammalian cells) and its partners Ufd1 and Npl4, in the export process across the ER membrane. Cdc48/p97 is able to participate in distinct cellular processes, membrane fusion or ubiquitin-dependent protein degradation for example, by interacting with different adaptor proteins specific for a particular process. In the case of ubiquitin-dependent protein degradation, the adaptor proteins are Ufd1 and Npl4.[62,63] The rate of degradation of misfolded proteins is significantly reduced in yeast having mutations in Cdc48, Ufd1 or Npl4.[64-68] Polyubiquitination, which occurs at the cytosolic surface of the ER membrane, has been shown necessary for the retro-translocation of some substrates.[59,66,69,70] However, polyubiquitination does not appear to be required in all instances of retro-translocation, since certain protein toxins that exploit the ERAD machinery to reach their substrates in the cytosol avoid ubiquitination altogether, partly to reduce the chance of being targeted to proteasomes,[71,72] and misfolded yeast pro-α-factor does not require ubiquitination.[73] From further study in mammalian cells, it has been proposed that polyubiquitin does not serve as a ratcheting molecule.[74] Instead polyubiquitin may serve as a recognition signal, possibly for the downstream-acting p97-Ufd1-Np14 complex that could be involved in either mobilising the dislocated substrate from other components in the membrane, in recruiting proteasomes to the membrane, or in the direct extraction of the protein from the membrane (reviewed in ref. 75). Release of at least some ubiquitinated substrates from the ER membrane to the cytosol occurs in a step that requires active proteasomes together with p97/Cdc48p and ATP.[76]

Whilst general features of ERAD are beginning to emerge, it is unlikely a single pathway will be involved. How a protein is selected and presented to the translocons for retro-translocation may vary between substrates. ERAD substrates may interact with different ER chaperones, some of which may be cell- or substrate-specific, and they may require the presence of different auxiliary proteins. Their retro-translocation requirements may or may not require ubiquitination or active proteasomes. Indeed, for some proteins, there may be no absolute requirement for ER translocons at all.[59] Whether retro-translocation occurs only from particular subcompartments of the ER is also unknown at present, and the precise role of vesicular transport in presenting substrates for retro-translocation is unclear. Considerable research effort is now required to elucidate these features and to provide detailed mechanisms into the process of retro-translocation, its gating and regulation.

References

1. Blobel G. Protein targeting. Chembiochem 2000; 1:86-102.
2. Johnson AE, van Maes MA. The translocon: a dynamic gateway at the ER membrane. Annu Rev Cell Develop Biol 1999; 15:799-842.
3. Rapoport TA, Jungnickel B, Kutay U. Protein transport across the eukaryotic endoplasmic reticulum and bacterial inner membranes. Annu Rev Biochem 1996; 65:271-303.
4. Helenius A, Marquardt T, Braakman I. The endoplasmic reticulum as a protein-folding compartment. Trends Cell Biol 1992; 2:227-31.
5. Ellgaard L, Helenius A. ER quality control: towards an understanding at the molecular level. Curr Opin Cell Biol 2001; 13:431-437.
6. Ellgaard L, Helenius A. Quality control in the endoplasmic reticulum. Nat Rev Mol Cell Biol 2003; 4:181-191.
7. Hong E, Davidson A, Kaiser C. A pathway for targeting soluble misfolded proteins to the yeast vacuole. J Cell Biol 1996; 135:623-633.
8. Klausner RD, Sitia R. Protein degradation in the endoplasmic reticulum. Cell 1990; 62:611-4
9. Brodsky JL, McCracken AA. ER protein quality control and proteasome-mediated protein degradation. Semin Cell Dev Biol 1999; 10:507-513.
10. McCracken AA, Brodsky JL. Assembly of ER-associated protein degradation in vitro: dependence on cytosol, calnexin, and ATP. J Cell Biol 1996; 132:291-8.
11. Thomas P, Qu B, Pedersen P. Defective protein folding as a basis of human disease. Trends Biochem Sci 1995; 20:456-459.
12. Dobson C. Protein folding and misfolding. Nature 2003; 426:884-890.

13. Kopito RR, Sitia R. Aggresomes and Russell bodies. Symptoms of cellular indigestion? EMBO Rep 2000; 1:225-31.
14. Shamu C, Cox J, Walter P. The unfolded protein response pathway in yeast. Trends Cell Biol 1998; 4:56-60.
15. Kozutsumi Y, Segal M, Gething M et al. The presence of malfolded proteins in the endoplasmic reticulum signals the induction of glucose-regulsted proteins. Nature 1988; 332:462-464.
16. Friedlander R, Jarosch E, Urban J et al. A regulatory link between ER-associated protein degradation and the unfolded-protein response. Nat Cell Biol 2000; 2:379-384.
17. Liu CY, Kaufman RJ. The unfolded protein response. J Cell Sci 2003; 116:1861-1862.
18. Bonifacino JS, Suzuki CK, Lippincott-Schwartz J et al. Pre-Golgi degradation of newly synthesized T-cell antigen receptor chains: intrinsic sensitivity and the role of subunit assembly. J Cell Biol 1989; 109:73-83.
19. Chen C, Bonifacino JS, Yuan LC et al. Selective degradation of T cell antigen receptor chains retained in a pre-Golgi compartment. J Cell Biol 1988; 107:2149-2161.
20. Lippincott-Schwartz J, Bonifacino JS, Yuan LC et al. Degradation from the endoplasmic reticulum: disposing of newly synthesized proteins. Cell 1988; 54:209-20.
21. Bonifacino JS, Lippincott-Schwartz J. Degradation of proteins within the endoplasmic reticulum. Curr Opin Cell Biol 1991; 3:592-600.
22. Ciechanover A. The ubiquitin-proteasome proteolytic pathway. Cell 1994; 79:13-21.
23. Sommer T, Jentsch S. A protein translocation defect linked to ubiquitin conjugation at the endoplasmic reticulum. Nature 1993; 365:176-9.
24. Ellgaard L, Molinari M, Helenius A. Setting the standards: quality control in the secretory pathway. Science 1999; 286:1882-1888.
25. Ward C, Omura S, Kopito R. Degradation of CFTR by the ubiquitin-proteasome pathway. Cell 1995; 83:121-127.
26. Jensen T, Loo M, Pind S et al. Multiple proteolytic systems, including the proteasome, contribute to CFTR processing. Cell 1995; 83:129-135.
27. Hampton RY. ER-associated degradation in protein quality control and cellular regulation. Curr Opin Cell Biol 2002; 14:476-82.
28. Ravid T, Doolman R, Avner R et al. The ubiquitin-proteasome pathway mediates the regulated degradation of 3-hydroxy-3-methylglutaryl-coenzyme A reductase. J Biol Chem 2000; 275:35840-35847.
29. Wiertz E, Jones T, Sun I et al. The human cytomegalovirus US11 gene product dislocates MHC class I heavy chains from the endoplasmic reticulum to the cytosol. Cell 1996; 84:769-779.
30. Wiertz EJ, Tortorella D, Bogyo M et al. Sec61-mediated transfer of a membrane protein from the endoplasmic reticulum to the proteasome for destruction. Nature 1996; 384:432-8.
31. Vashist S, Kim W, Belden WJ et al. Distinct retrieval and retention mechanisms are required for the quality control of endoplasmic reticulum protein folding. J Cell Biol 2001; 155:355-68.
32. Haynes CM, Caldwell S, Cooper AA. An HRD/DER-independent ER quality control mechanism involves Rsp5p-dependent ubiquitination and ER-Golgi transport. J Cell Biol 2002; 158:91-101.
33. Taxis C, Vogel F, Wolf DH. ER-Golgi traffic is a prerequisite for efficient ER degradation. Mol Biol Cell 2002; 13:1806-18.
34. Sato M, Sato K, Nakano A. Endoplasmic reticulum quality control of unassembled iron transporter depends on Rer1p-mediated retrieeval from the Golgi. Mol Biol Cell 2004; 15:1417-1424.
35. Kamhi-Nesher S, Shenkman M, Tolchinsky S et al. A novel quality control compartment derived from the endoplasmic reticulum. Mol Biol Cell 2001; 12:1711-23.
36. Caldwell SR, Hill KJ, Cooper AA. Degradation of endoplasmic reticulum (ER) quality control substrates requires transport between the ER and Golgi. J Biol Chem 2001; 276:23296-23303.
37. Vashist S, Ng D. Misfolded proteins are sorted by a sequential checkpoint mechanism of ER quality control. J Cell Biol 2004; 165:41-52.
38. Molinari M, Helenius A. Chaperone selection during glycoprotein translocation into the endoplasmic reticulum. Science 2000; 288:331-3.
39. Molinari M, Galli C, Piccaluga V et al. Sequential assistance of molecular chaperones and transient formation of covalent complexes during protein degradation from the ER. J Cell Biol 2002; 158:247-57.

40. Zhang JX, Braakman I, Matlack KE et al. Quality control in the secretory pathway: the role of calreticulin, calnexin and BiP in the retention of glycoproteins with C-terminal truncations. Mol Biol Cell 1997; 8:1943-54.
41. Zhang Y, Nijbroek G, Sullivan ML et al. Hsp70 molecular chaperone facilitates endoplasmic reticulum-associated protein degradation of cystic fibrosis transmembrane conductance regulator in yeast. Mol Biol Cell 2001; 12:1303-14.
42. Nishikawa SI, Fewell SW, Kato Y et al. Molecular chaperones in the yeast endoplasmic reticulum maintain the solubility of proteins for retrotranslocation and degradation. J Cell Biol 2001; 153:1061-70.
43. Caramelo JJ, Castro OA, Alonso LG et al. UDP-Glc:glycoprotein glucosyltransferase recognises structured and solvent accessible hydrophobic patches in molten globule-like folding intermediates. Proc Natl Acad Sci USA 2003; 100:86-89.
44. Parodi AJ. Protein glucosylation and its role in protein folding. Annu Rev Biochem 2000; 69:69-93.
45. Oda Y, Hosokawa N, Wada I et al. EDEM as an acceptor of terminally misfolded glycoproteins released from calnexin. Science 2003;299:1394-7.
46. Molinari M, Calanca V, Galli C et al. Role of EDEM in the release of misfolded glycoproteins from the calnexin cycle. Science 2003; 299:1397-400.
47. Jakob C, Burda P, Roth J et al. Degradation of misfolded endoplasmic reticulum glycoproteins in Saccharomyces cerevisiae is determined by a specific oligosaccharide structure. J Cell Biol 1998; 142:1223-1233.
48. Tokunaga F, Brostrom C, Koide T et al. Endoplasmic reticulum (ER)-associated degradation of misfolded N-linked glycoproteins is suppressed upon inhibition of ER mannosidase I. J Biol Chem 2000; 275:40757-64.
49. Knittler MR, Dirks S, Haas IG. Molecular chaperones involved in protein degradation in the endoplasmic reticulum: quantitative interaction of the heat shock cognate protein BiP with partially folded immunoglobulin light chains that are degraded in the endoplasmic reticulum. Proc Natl Acad Sci USA 1995; 92:1764-8.
50. Knittler MR, Haas IG. Interaction of BiP with newly synthesized immunoglobulin light chain molecules: cycles of sequential binding and release. EMBO J 1992; 11:1573-81.
51. Brodsky JL, Werner ED, Dubas ME et al. The requirement for molecular chaperones during endoplasmic reticulum-associated protein degradation demonstrates that protein export and import are mechanistically distinct. J Biol Chem 1999; 274:3453-3460.
52. Plemper RK, Bohmler S, Bordallo J et al. Mutant analysis links the translocon and BiP to retrograde protein transport for ER degradation. Nature 1997; 388:891-5
53. Gilbert H. Protein disulfide isomerase and assisted protein folding. Biol Chem 1997; 272:29399-29402.
54. Gillece P, Luz JM, Lennarz WJ et al. Export of a cysteine-free misfolded secretory protein from the endoplasmic reticulum for degradation requires interaction with protein disulfide isomerase. J Cell Biol 1999; 147:1443-56.
55. Fagioli C, Mezghrani A, Sitia R. Reduction of interchain disulfide bonds precedes the dislocation of Ig-mu chains from the endoplasmic reticulum to the cytosol for proteasomal degradation. J Biol Chem 2001; 276:40962-40967.
56. Wang Q, Chang A. Eps1, a novel PDI-related protein involved in ER quality control in yeast. EMBO J 1999; 18:5972-82.
57. Wang Q, Chang A. Substrate recognition in ER-associated degradation mediated by Eps1, a member of the protein disulfide isomerase family. EMBO J 2003; 22:3792-802.
58. Pilon M, Schekman R, Romisch K. Sec61p mediates export of a misfolded secretory protein from the endoplasmic reticulum to the cytosol for degradation. EMBO J 1997; 16:4540-8.
59. Biederer T, Volkwein C, Sommer T. Degradation of subunits of the Sec61p complex, an integral component of the ER membrane, by the ubiquitin-proteasome pathway. EMBO J 1996; 15:2069-2076.
60. Hamman BD, Chen J-C, Johnson EE et al. The aqueous pore through the translocon has a diameter of 40-60Å during cotraanslational protein translocation at the ER membrane. Cell 1997; 89:535-544.

61. Kosova Z, Wolf DH. For whom the bell tolls: protein quality control of the endoplasmic reticulum and the ubiquitin-proteasome connection. EMBO J 2003; 22:2309-2317.
62. Meyer H, Shorter J, Seemann J et al. A complex of mammalian Ufd1 and Npl4 links the AAA ATPase , p97, to ubiquitin and nuclear transport pathways. EMBO J 2000; 19:2181-2192.
63. Hitchcock A, Krebber H, Frietze S et al. The conserved Npl4 protein complex mediates proteasomal-dependent membrane-boundtranscription factor activation. Mol Biol Cell 2001; 12:3226-3241.
64. Ye Y, Meyer HH, Rapoport TA. The AAA ATPase Cdc48/p97 and its partners transport proteins from the ER into the cytosol. Nature 2001; 414:652-6.
65. Bays N, Wilhovsky S, Goradia A et al. HRD4/NPL4 is required for the proteasomal processing of ubiquitinated ER proteins. Mol Biol Cell 2001; 12:4114-4128.
66. Jarosch E, Taxis C, Volkwein C et al. Protein dislocation from the ER requires polyubiquitination and the AAA-ATPase Cdc48. Nat Cell Biol 2002; 4:134-9
67. Rabinovich E, Kerem A, Frohlich K-U et al. AAA-ATPase p97/Cdc48p, a cytosolic chaperone required of endoplasmic reticulum-associated protein degradation. Mol Cell Biol 2002; 22:626-634.
68. Braun S, Matuschewski K, Rape M et al. Role of trhe ubiquitin-selective $CDC48^{UFD1/NPL4}$ chaperone (segregase) in ERAD of OLE1 and other substrates. EMBO J 2002; 21:615-621.
69. de Virgilio M, Weninger H, Ivessa NE. Ubiquitination is required for the retro-translocation of a short-lived luminal endoplasmic reticulum glycoprotein to the cytosol for degradation by the proteasome. J Biol Chem 1998; 273:9734-9743
70. Kikkert M, Hassink G, Barel M et al. Ubiquitination is essentila for human cytomegalovirus US11-mediated dislocation of MHC class 1 molecules from the endoplasmic reticulum to the cytosol. Biochem J 2001; 358:369-377.
71. Deeks ED, Cook JP, Day PJ et al. The low lysine content of ricin A chain reduces the risk of proteolytic degradation after translocation from the endoplasmic reticulum to the cytosol. Biochemistry. 2002; 41:3405-3413.
72. Rodighiero C, Tsai B, Rapoport TA et al. Role of ubiquitination in retro-translocation of cholera toxin and escape of cytosolic degradation. EMBO Rep 2002;3:1222-7.
73. Werner ED, Brodsky JL, McCracken AA. Proteasome-dependent endoplasmic reticulum-associated protein degradation: an unconventional route to a familiar fate. Proc Natl Acad Sci USA 1996; 93:13797-801.
74. Flierman D, Ye Y, Dai M et al. Polyubiquitin serves as a recognition signal, rather than a ratcheting molecule, during retrotranslocation of proteins across the endoplasmic reticulum membrane. J Biol Chem 2003; 278:34774-34782.
75. Jarosch E, Geiss-Friedlander R, Meusser B et al. Protein dislocation from the endoplasmic reticulum-pulling out the suspect. Traffic 2002; 3:530-6.
76. Elkabetz Y, Shapira H, Rabinovich E et al. Distinct steps in the dislocation of luminal endoplasmic reticulum-associated degradation substrates. Roles of endoplasmic reticulum-boundp97/Cdc48p and proteasome. J Biol Chem 2004; 279:3980-3989.

CHAPTER 8

Chloroplast Protein Targeting:

Multiple Pathways for a Complex Organelle

Matthew D. Smith and Danny J. Schnell*

Abstract

Plastids, exemplified by chloroplasts, are a diverse group of essential organelles that distinguish plant cells. The biogenesis of these organelles is essential to plant growth and development, and relies on the import of >2500 nuclear-encoded proteins from the cytoplasm. The import of the large majority of these proteins is dependent on the Toc-Tic machinery of the chloroplast envelope. However, an ever increasing number of new pathways for targeting proteins to numerous chloroplast sub-compartments have been identified. Furthermore, it appears that the multiple targeting pathways and the regulation of import play direct roles in the differentiation and specific functions of distinct plastid types during plant growth and development. This chapter summarizes the state of the field, emphasizing the mechanisms of targeting proteins to and across the plastid envelope, and to chloroplast sub-compartments.

Introduction

The plastids correspond to a diverse group of essential organelles that are a distinguishing feature of plant cells.[1] The most familiar type of plastid is the chloroplast, which houses the photosynthetic apparatus in green tissues. However, plastids are present in all plant cells and are the site of other equally important biochemical processes, including crucial steps in lipid, amino acid, starch, nitrogen and sulphur metabolism.[2] Therefore, the biogenesis and maintenance of plastids in all tissues is essential to the growth and development of plants.

Plastids are widely accepted to have originated from an endosymbiotic event when a primitive photosynthetic cyanobacterium was engulfed and maintained by a nucleated and mitochondriate cell.[3] As part of the evolutionary transition from free-living cyanobacterium to sub-cellular organelle, the majority of plastid genes were transferred to the nuclear genome. The resulting genome of present-day semi-autonomous plastids in vascular plants encodes only ~120 genes which are primarily involved in photosynthesis, and transcription and translation of the plastid-encoded genes.[4] Accordingly, the vast majority of plastid proteins (estimates range from 2500 to 3800 distinct proteins, accounting for ~95% of all plastid proteins) are encoded in the nucleus and translated in the cytoplasm.[5,6] To compensate for the massive gene

*Corresponding Author: Danny J. Schnell—Department of Biochemistry and Molecular Biology, University of Massachusetts, Amherst, Massachusetts 01003, U.S.A.
Email: dschnell@nsm.umass.edu

Protein Movement Across Membranes, edited by Jerry Eichler. ©2005 Eurekah.com and Springer Science+Business Media.

transfer, plastids have evolved a highly accurate protein trafficking system for identifying these proteins in the cytoplasm, targeting them to the plastid surface and translocating them across multiple distinct membranes to their final residence within the organelle.[7] This review will focus on the general mechanisms of protein targeting to and across plastid membranes and will highlight the recent discovery of the role of the import machinery during plastid development and differentiation. We will emphasize the mechanisms of targeting proteins to and across the plastid envelope and briefly summarize the pathways for protein targeting to the thylakoid membrane.

General Features of Plastid Protein Import

All plastids are surrounded by a double membrane envelope, which serves to separate the internal stroma from the cytoplasm (Fig. 1). In addition, several plastid types contain internal membrane systems, such as the thylakoid in chloroplasts. This complex architecture requires trafficking systems to target proteins to at least six suborganellar compartments; the outer and inner envelope membranes, the intermembrane space of the envelope, the stroma, and the thylakoid membrane and lumen (Fig. 1). The intricate structure of plastids requires that many

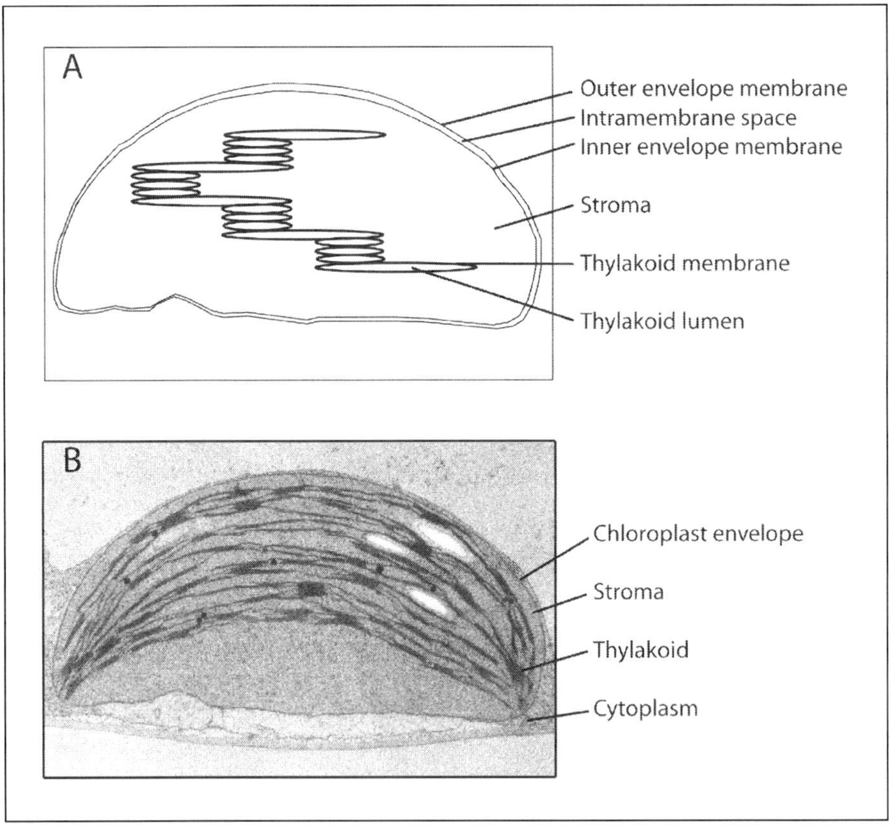

Figure 1. Morphology of chloroplasts. A) diagrammatic representation of a chloroplast illustrating the six sub-compartments of the organelle. B) transmission electron micrograph of an *Arabidopsis thaliana* leaf chloroplast.

proteins contain compound targeting signals that are read sequentially as the protein is targeted first to the envelope and subsequently to a final suborganellar location. The machinery for targeting to and across the envelope membranes appears to be unique to plastids and likely evolved in response to endosymbiosis.[7] In contrast, the trafficking systems of the thylakoid are conserved from their bacterial origin and have been adapted to target both nuclear-encoded and plastid-encoded polypeptides.[8]

The complexity of plastid trafficking systems is compounded by the dual genetic origin of the organelle and the considerable biochemical and morphological changes that occur during plastid development in different plant tissues. For example, many plastid protein complexes, in particular those involved in plastid genome replication, plastid gene expression and photosynthesis, contain subunits encoded by both nuclear and plastid genomes. As a consequence, the import and assembly of these hybrid complexes must be tightly coordinated. In addition, the levels and compositions of imported proteins vary dramatically during plant development and in different plastid types. The most notable example is the development of chloroplasts from undifferentiated proplastids during plant greening (photomorphogenesis). In this instance, the expression of genes encoding photosynthetic proteins increases several orders of magnitude,[9] and the protein trafficking system must adapt to accommodate the shift from a relatively low capacity system for targeting constitutively-expressed plastid proteins to one that can accommodate the massive influx of specific photosynthetic cargo. Although this diversity of plastid form and function and their ability to undergo such dramatic interconversion is controlled by the nuclear genome,[9] recent evidence indicates that the import machinery simply does not serve a housekeeping function, but rather plays an active role in regulating plastid biogenesis and differentiation.

Most Proteins Are Targeted to Plastids via Cleavable Transit Peptides

The vast majority of nuclear-encoded plastid proteins are translated in the cytoplasm as preproteins, containing cleavable N-terminal extensions called transit peptides. Transit peptides are the signals that mark these proteins for delivery to plastids.[10] With rare exception, cleavable transit peptides are found on nuclear-encoded preproteins destined for all internal compartments of plastids (i.e., all compartments other than the outer envelope membrane). The majority of transit peptides can be classified as stromal transfer signals, as the preproteins of which they are a part are translocated across the double membrane envelope and delivered to the stroma (Table 1). Despite the functional similarity in stromal transfer signals, they lack consensus structural features that clearly define them as transit peptides.[11] Transit peptides can vary from thirty to well over one hundred amino acids in length.[10] In general, they tend to be devoid of acidic residues and enriched in hydroxylated amino acids, resulting in a tendency to overall basic charge. In contrast to many mitochondrial presequences, transit peptides do not form amphipathic helices. Nonetheless, a number of algorithms have been developed to predict the localization of nuclear-encoded plastid proteins based on the characteristics of their N-terminal sequences.[12-14] Upon emerging into the stroma, the transit peptide is recognized and cleaved by the stromal processing peptidase,[15,16] yielding a protein that can then be folded and/or assembled into a functional complex, or subsequently targeted to one of the other sub-chloroplast compartments.

Molecular machinery located at the envelope membranes of chloroplasts is responsible for recognizing and translocating preproteins post-translationally from the cytoplasm into the organelle in a process that requires both ATP and GTP.[17-19] In contrast to mitochondria, a membrane potential is not required for envelope translocation. Over the last decade, numerous components of the plastid preprotein import machinery have been identified (Fig. 2). The translocon at the outer envelope membrane of chloroplasts (Toc complex)[20] is responsible for

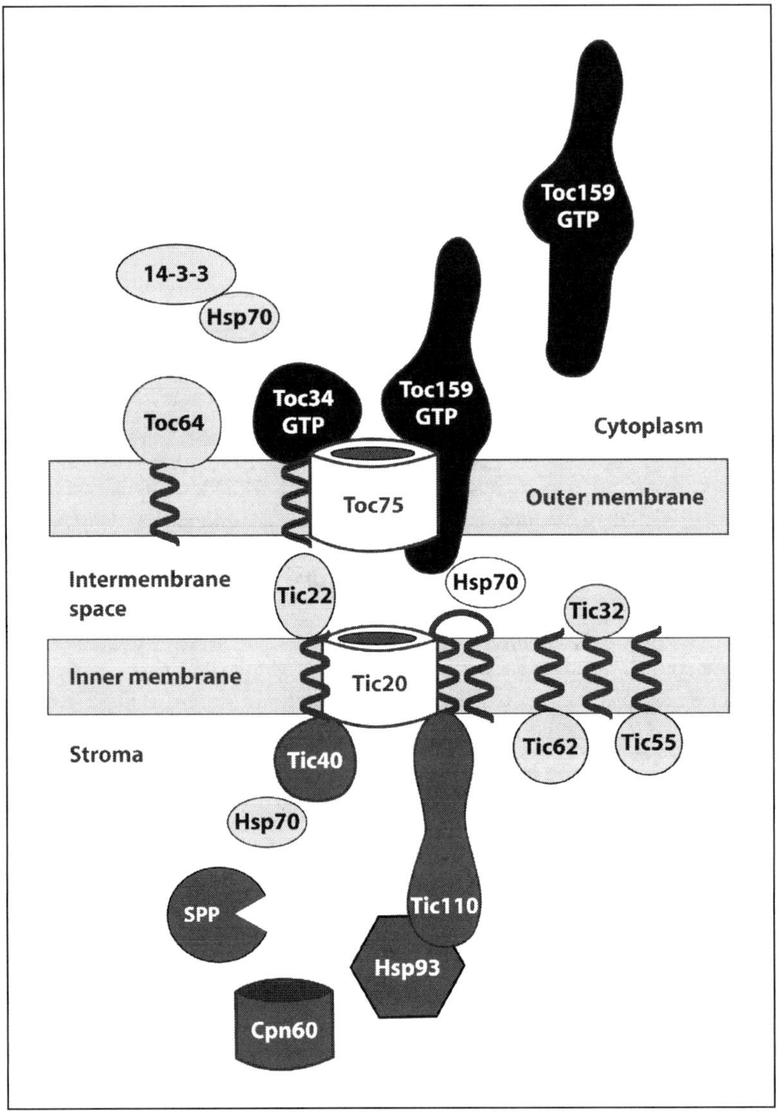

Figure 2. Schematic representation of components of the Toc and Tic translocon complexes involved in the recognition and import of preproteins from the cytoplasm into the stroma. These include core translocon components implicated in preprotein targeting (black), membrane translocation (white) and the translocation and maturation of preproteins in the stroma (dark gray). Additional proteins with unknown functions or proposed roles in facilitating or regulating import under specialized circumstances also are shown (light gray).

recognizing plastid preproteins and initiating protein import into plastids. It acts coordinately with the Tic (translocon at the inner envelope membrane of chloroplasts) complex to complete translocation of preproteins from the cytoplasm across the envelope and into the stroma, thereby bypassing the possibility of mis-targeting to the intermembrane space.

Table 1. The architecture of preproteins and the multiple pathways for targeting to chloroplast sub-compartments

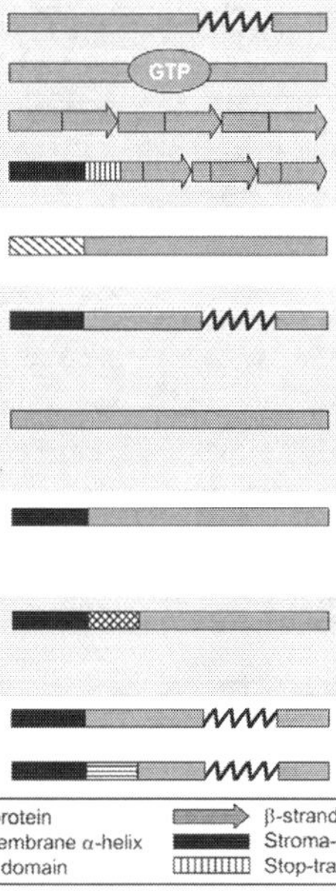

Chloroplast Compartment	Preprotein Architecture	Targeting components/ Pathway used	Examples
Outer Membrane	A.	Toc75 or spontaneous	OEP14, Toc34, DGD1
	B.	Toc34, Toc75	Toc159
	C.	Unknown proteinaceous factors	OEP80
	D.	Toc-Tic pathway	Toc75
Intermembrane Space		Unknown proteinaceous factors	Tic22
Inner Membrane	A.	1. Toc-Tic pathway, via stroma	Tic110
		2. Toc-Tic pathway, lateral transfer within IM	TPT
	B.	1. Unknown proteinaceous factors, via stroma	ceQORH
		2. Unknown proteinaceous factors, via IMS	Tic32/IEP32
Stroma		1. Toc-Tic pathway	SSU, Fd, E1α
		2. Toc-Tic pathway or Pchlide-regulated Ptc?	PORA
Thylakoid lumen		1. ATP-dependent cpSec pathway	PC, OE33
		2. Twin-arginine dependent TAT/ΔpH pathway	OE17, OE23
Thylakoid membrane	A.	GTP-dependent cpSRP pathway	LHCP
	B.	"Spontaneous" insertion	CFo subunit II

Mature protein; Trans-membrane α-helix; GTPase domain; β-strand; Stroma-targeting transit peptide; Stop-transfer signal; IMS targeting signal; cpSec or TAT signal; Sec-like "spontaneous" signal

Initial Targeting and Translocation of Preproteins Is Mediated by the Toc Complex

Four proteins have been identified in the Toc complex (Fig. 2). The essential core of the complex consists of two membrane-associated GTPases, Toc34 and Toc159, and a β-barrel membrane channel, Toc75, with the numbers representing the molecular size of the proteins in kDa.[21] These three components assemble into a large toroid-shaped structure of >500 kDa to form the functional translocon.[22] Toc34 and Toc159 mediate the initial binding of preproteins at the chloroplast surface via direct interactions with transit peptides.[23-27] GTP binding and/or hydrolysis at the GTPases are required to initiate translocation of the preprotein across the outer membrane.[28,29] Toc75 comprises a major component of the protein-conducting channel,[26,30,31] and along with Toc159, appears to mediate outer membrane translocation.[25,32] Toc64 is the fourth potential protein of the Toc complex,[33] however, it does not appear to be stably associated with the other Toc components,[22] and it is not required for import of plastid preproteins in a reconstituted system.[35] Toc64 homologues also are found in other sub-cellular compartments.[34] Therefore, it is not yet clear what role Toc64 might play in plastid protein import.

The GTP-binding domains (G domains) of Toc34 and Toc159 are homologous, and together, they define a unique subclass of GTPase that is distantly related to the classical Ras GTPase superfamily.[36] Null mutants lacking the Toc159 or Toc34 families (see below) of GTPases are inviable,[37-39] indicating that both GTPases are essential for plastid biogenesis. The role of GTP and the mechanism by which the Toc GTPases control the initial stages of preprotein targeting remain areas of intense investigation.[40] Two major activities have been proposed. First, Toc34 and Toc159 are proposed to form a gate that regulates access of preproteins to the translocation channel of the Toc complex. Considerable evidence suggests that GTP-regulated dimerization of the two GTPases underlies the mechanism of the gate.[41-45] In this scenario, the Toc34-Toc159 heterodimer represents a closed gate that prevents access of inappropriate cytoplasmic proteins to the Toc translocation channel.[40] The binding of a preprotein containing an authentic transit peptide at the receptor sites on the Toc GTPases results in dissociation of the heterodimer and transfer of the preprotein from the initial receptor site into the Toc channel containing Toc75. The second proposed function of GTP is to provide the initial energy for insertion of the preprotein across the outer membrane.[40] Alternative models have been proposed to account for the role of GTP at this stage. One model suggests that GTP binding at the GTPases is sufficient to transfer the preprotein into the channel.[28] The binding of an Hsp70-type molecular chaperone in the intermembrane space would provide the energy for translocation across the membrane. The chaperone would account for the ATP requirement of outer membrane translocation.[46] The second model suggests that Toc159 acts as a repetitive GTP-driven motor with repeated rounds of GTP hydrolysis driving outer membrane translocation.[35] This model is based on the observation that Toc159 and Toc75 are sufficient to catalyze at least partial translocation of a preprotein in reconstituted proteoliposomes. However, GTP hydrolysis alone is insufficient to drive outer membrane translocation in isolated chloroplasts, a process that requires low levels of ATP hydrolysis in the intermembrane space.[29,46] While there is a clear consensus that the activities of the GTPases are critical in mediating the recognition of preproteins and regulating translocation, the exact mechanism by which GTP functions at the early stages of import remains to be precisely defined.

Although Toc34 and Toc159 are the primary docking sites for preproteins at the chloroplast surface, several cytoplasmic factors also have been implicated in the targeting reaction (Fig. 2). A guidance complex has been identified in wheat germ extracts that appears to stimulate preprotein import into isolated chloroplasts. The complex contains a cytoplasmic Hsp70-type chaperone and a 14-3-3 protein.[47] The Hsp70 chaperone presumably assists in maintaining preproteins in an unfolded state that is competent for membrane translocation,[48] although this has recently been challenged.[49] 14-3-3 proteins are typically associated with kinase activities, leading to the proposal that a phosphorylation cycle might be involved in the targeting reaction.[24,47,50,51] However, phosphorylation of import components or preproteins is not required for protein import in vitro or in vivo,[48,52] leaving the physiological role of the guidance complex unresolved. A cytoplasmic form of Toc159 also has been detected.[53-55] This observation, coupled with the documented role of Toc159 GTPase activity in targeting it to the membrane,[44,54,55] has led to the proposal that Toc159 might function as a cycling receptor by binding to preproteins in the cytoplasm and delivering them to the Toc complex. To date, there is no direct evidence for such a function, and the nature of the soluble form of Toc159 remains under investigation.

Unlike Toc159 and Toc34, Toc75 has apparently been adapted from β-barrel membrane proteins that are commonly found in the outer membranes of Gram-negative bacteria, including cyanobacteria.[7,56] Toc75 can form a membrane channel with a conductance sufficient to accommodate an unfolded polypeptide, supporting its role as a component of the Toc translocation channel.[31,57] Early import intermediates that are trapped at the stage of outer membrane translocation crosslink both to Toc75 and to the membrane anchor domain of Toc159,[25,32] suggesting that both proteins are required to form a functional translocation channel. The participation of Toc159 in membrane translocation provides a possible mechanism of direct coupling of GTP-regulated preprotein recognition with membrane translocation.

The Toc and Tic Translocons Cooperate to Mediate Preprotein Transport from the Cytoplasm to the Stroma

Upon emerging from the Toc complex, the preprotein is immediately engaged by components of the Tic translocon. Unlike the Toc translocon, which appears to be a relatively stable membrane complex, the Tic translocon is dynamic.[58] Tic components assemble in response to preprotein translocation by physically associating with the Toc complex to form functional import sites, referred to as supercomplexes, at envelope contact sites.[58,59] Translocation across the inner membrane requires only stromal ATP hydrolysis.[60,61] As with outer membrane translocation, the ATP requirement is attributed to molecular chaperones located on the trans side of the membrane, in this case corresponding to the stroma.[48] Although not directly demonstrated in plastids, the chaperones are proposed to bind preproteins and thereby drive unidirectional transport into the stroma in a manner analogous to that proposed for the mitochondrial Hsp70.[62]

Four polypeptides of the inner membrane directly participate in import, as demonstrated by their ability to covalently cross-link to preproteins during inner membrane translocation (Fig. 2). Tic22 is a resident of the intermembrane space, and this localization suggests that it might function in the assembly of the Tic complex or the Toc-Tic supercomplex.[58] Tic20 and Tic110 both have been implicated in the membrane translocation reaction.[63,64] Tic20 is distantly related to bacterial branched-chain amino acid transporters and to the Tim17/23 components of the mitochondrial inner membrane translocase.[58,65] It is a polytopic integral membrane protein that interacts with preproteins during translocation.[58] In addition, anti-sense down regulation of Tic20 results in a specific defect in transport across the inner membrane.[63]

Tic110 is an abundant inner membrane protein and a fraction of it is found in association with Toc components under steady-state conditions, indicating a central role for this component in Tic complex function.[59,66] In vitro analysis of Tic110 has led to the proposal that it coordinates the late events in preprotein import.[59,67] The ~95 kDa stromal domain of the protein possesses two critical activities. First, it contains a transit peptide-binding site adjacent to its membrane anchor segments.[67] This site is proposed to form the initial binding site for the preprotein as it emerges from the Tic channel, thereby preventing it from slipping back into the intermembrane space. Tic110 also specifically associates with the stromal Hsp93 chaperone, and it is believed that the chaperone binds to the preprotein and provides the driving force for subsequent translocation.[59,68] Tic40 is a third integral membrane component of Tic complexes. Although its role in import is not essential, the fact that it possesses a domain similar to several cochaperone molecules suggests that it might play a role in coordinating the association of chaperones with preproteins during the late stages of import.[69] In addition to assisting in translocation, the molecular chaperones likely facilitate folding of newly imported proteins in the stroma. Cpn60, the plastid GroEL homologue, also associates with import complexes,[66] suggesting coordination between preprotein translocation, processing and folding. Stromal Hsp70 has not been shown to directly participate in the import reaction, but it does associate with some nuclear-encoded thylakoid proteins to assist in their transit through the stroma from the envelope translocons to the thylakoid membrane.[70]

At least three other proteins have been implicated in preprotein translocation across the inner membrane (Fig. 2). Tic62, Tic55 and Tic32/IEP32 have all been identified as potentially being complexed with Tic110 and have been proposed to play regulatory roles in import.[71-74] While Tic32 appears to be an essential protein in Arabidopsis,[73] direct evidence for their roles in protein import or their regulation is still lacking for all three of these proteins.

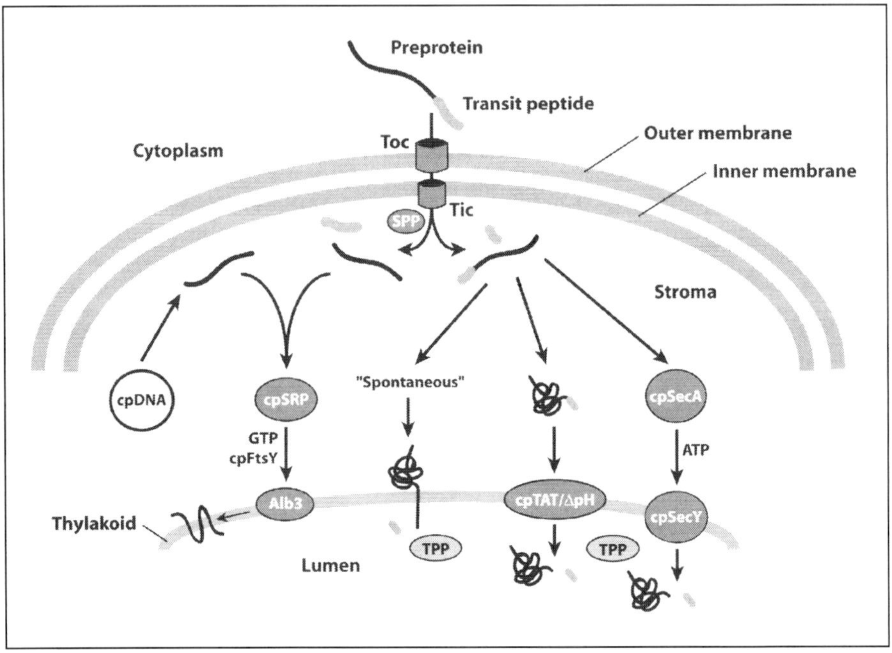

Figure 3. Schematic representation of the four known pathways of protein targeting to the thylakoid within chloroplasts. See text for details. SPP, stromal processing peptidase, TPP, thylakoid processing peptidase.

Thylakoid Proteins Are Targeted by Pathways Conserved from Prokaryotes

Nuclear genes encode approximately 50% of thylakoid polypeptides. The process of protein trafficking to the thylakoid is complex, comprising four different pathways (Fig. 3). As previously mentioned, these pathways all appear to be conserved from the secretion and insertion pathways that exist at the cytoplasmic membrane of prokaryotes.[8] Thus, it appears that the conserved pathways of protein targeting were retained within plastids, but relocated to the thylakoid to maintain proper protein topogenesis as gene transfer occurred following endosymbiosis. Thylakoid biogenesis warrants an entire review on its own, and we will only summarize the salient features of these pathways in this review.

All nuclear-encoded thylakoid proteins are first targeted to the stroma using transit peptides and the Toc-Tic machinery (Table 1, Fig. 3). The transit peptides of proteins destined for the thylakoid lumen are followed in tandem by lumenal targeting domains that are functionally interchangeable with the signal sequences that target bacterial proteins for secretion (Table 1).[75] These signal sequences are ~35 residues in length and are characterized by a hydrophobic core (~15 amino acids) flanked by relatively hydrophilic regions.[76,77] Despite this consensus, lumenal targeting signals can be divided into two classes, representing two distinct targeting pathways. The two classes are distinguished by the presence of a twin arginine motif that immediately precedes the hydrophobic core of the signal sequence.[78] Proteins lacking this motif utilize the cpSec-dependent pathway for targeting. Components of this pathway include orthologues of the soluble SecA targeting receptor and the SecY membrane translocon of Bacteria.[79,80] The cpSec pathway is driven by ATP hydrolysis and is utilized by proteins that are transported in an unfolded conformation (Fig. 3). Preproteins containing signal sequences with the twin arginine motif utilize a pathway termed the cpTAT or Δ-pH pathway. The Hcf106, Tha4 and cpTatC components of this pathway are comparable to the TatABC system of prokaryotes.[81-83] The cpTAT pathway is distinguished by its ability to transport fully folded and/or oligomeric proteins using the pH gradient across the thylakoid as the sole energy source (Fig. 3). It apparently evolved to handle the transport of proteins that require folding and the catalyzed attachment of cofactors in the stroma or bacterial cytoplasm prior to transport.[84]

The two remaining thylakoid targeting pathways are specific for integral thylakoid membrane proteins (Table 1). The cpSRP pathway is orthologous to the SRP-dependent pathways of Bacteria and the eukaryotic ER.[85] Unlike the bacterial and ER SRP molecules, cpSRP does not contain an RNA component.[85,86] The major substrates for this pathway are the light harvesting chlorophyll binding proteins (LHCPs) that form the antennae of the photoreaction centers. Segments comprised of at least one transmembrane domain target these polytopic membrane proteins to the thylakoid.[87] The signal is bound by the stromal cpSRP, and the proteins are targeted to the thylakoid via a GTP-regulated cognate interaction between cpSRP and its receptor at the membrane, cpFtsY (Fig. 3).[86] Integration of LHCPs requires ALB3, an integral thylakoid membrane protein that is related to the Oxa1p and YidC proteins that operate in mitochondria and Bacteria, respectively.[88,89] The cpSRP pathway is also involved in the targeting of plastid-encoded thylakoid membrane proteins.[90] In this case, cpSRP acts cotranslationally to target nascent chain-ribosome complexes to the membrane in a manner analogous to the eukaryotic SRP that mediates cotranslational targeting to the eukaryotic endoplasmic reticulum (Fig. 3). The final pathway for thylakoid targeting is referred to as the 'spontaneous' pathway because it does not require a measurable energy input or proteinaceous components in the stroma or at the membrane.[91] The substrates for this pathway include subunit II of the chloroplast F_0 ATP synthase and several photosystem II proteins.[92-94] Subunit II is synthesized with a bipartite targeting signal which is similar to those used in the cpSec and cpTat/ΔpH pathways. As is the case for the proteins targeted by the cpSec and cpTat/ΔpH pathways, the second part of the targeting signal for proteins inserted using the spontaneous pathway is cleaved by the thylakoid processing peptidase in the lumen (Fig. 3).

Inner Envelope Membrane Proteins May Use Distinct Mechanisms of Insertion

Our knowledge of the mechanism of targeting to the inner envelope membrane is based on studies of a very few proteins. The majority of these proteins contain a stromal-targeting transit peptide that is cleaved by the stromal processing peptidase (Table 1). The proteins are subsequently targeted to the inner membrane by signals contained within the mature portions of the proteins once the transit peptide has been removed. The signal for targeting to the inner membrane resides within one or more of the transmembrane domains of these proteins. Fusion proteins containing the transit peptide and transmembrane segments of Tic110 can be captured as stromal intermediates en route to the inner membrane, suggesting that the protein might first enter the stroma on the Toc-Tic pathway and then insert into the inner membrane from the stromal side.[95] In contrast, no stromal intermediates are detected during the targeting of the TPT transporter, a polytopic inner membrane carrier protein.[96] This result suggests that the protein is integrated into the inner membrane by lateral diffusion of its transmembrane segments into the bilayer from the Tic translocon. The inner envelope protein IEP32/Tic32 has recently been proposed to use yet another pathway for targeting to the inner membrane. This protein appears to be targeted to chloroplasts by a novel mechanism that doesn't involve the Toc-Tic pathway or the use of a cleavable transit peptide, and may be inserted into the inner membrane directly from the intermembrane space with the assistance of Tic22.[74] It is therefore possible that integration into the inner membrane might involve distinct mechanisms depending on the substrate. In mitochondria, polytopic carrier proteins are integrated via an inner membrane translocase separate from that utilized by matrix proteins or other inner membrane proteins.[97] The components that decode the signals for inner membrane integration have not been identified in plastids. Therefore, it remains to be seen whether the stromal, intermembrane space and lateral diffusion mechanisms truly exist and whether components distinct from the known Tic components are involved in targeting to the inner membrane.

Outer Membrane Proteins Contain Intrinsic Uncleaved Targeting Signals

The targeting of several outer membrane proteins have been studied, including the three core components of the Toc complex, Toc159, Toc75 and Toc34, as well OEP14 (OEP7), a protein with a single α-helical membrane segment. With the exception of Toc75, these proteins do not contain cleavable N-terminal targeting signals, but rather contain intrinsic targeting information within their mature portions (Table 1).

OEP14 is a small protein of chloroplast outer envelope membranes that has been the subject of a number of studies on targeting to the outer membrane. The intrinsic targeting information is contained within its first 30 amino acids, which includes its transmembrane segment.[98] Although OEP14 insertion does not require exogenous energy,[99] a recent study demonstrates that Toc75 mediates targeting and insertion of OEP14.[100] The insertion of Toc34 and an additional outer membrane protein, DGD1, are also directed by signals encompassing transmembrane segments.[101-103] Although their insertion is not sensitive to pretreatment of chloroplasts with protease, it is possible that both of these proteins also utilize Toc75 for insertion because Toc75 is known to be protease resistant. GTP stimulates Toc34 insertion, presumably because nucleotide binding by the GTPase stabilizes a conformation that is more import competent.[101,104]

Toc159 also is targeted to the outer envelope without a cleavable targeting signal. Targeting occurs in two distinct stages. Specific binding at the chloroplast surface appears to be mediated by its GTPase domain through an interaction with the homologous GTPase domain of Toc34.[44,54,105] Insertion of the membrane-anchor domain of Toc159 into the outer membrane is a distinguishable step from binding that requires an interaction with Toc75.[44,105] Unlike the previously mentioned outer membrane proteins, Toc159 does not contain an α-helical transmembrane segment.

The identification of Toc75 as a component of multiple outer membrane insertion and translocation pathways suggests that it might provide a general pore for access into or across the outer membrane. In this scenario, association of Toc75 with different regulators (e.g., the Toc GTPases) would dictate the specificity of the pathway. It remains to be conclusively demonstrated whether the targeting pathways for outer membrane proteins and proteins destined for the plastid interior converge at Toc75 channels that are part of a translocon complex or whether outer membrane proteins utilize a pool of Toc75 in the outer envelope that is not associated with other Toc components.[100] A significant fraction of Toc75 is not associated with the Toc GTPases,[58] meaning sufficient additional Toc75 is available to mediate other targeting processes.

The targeting pathway used to direct Toc75 itself to the outer membrane is unique among outer envelope membrane proteins. It is the only known example that is directed to the outer envelope membrane by a unique, cleavable and bipartite, N-terminal transit peptide.[106] The N-terminal part of the transit peptide directs the protein to the stroma via the Toc-Tic machinery and is cleaved by the stromal processing peptidase,[107] whereas the C-terminal part contains a poly-glycine motif that acts as a 'stop-transfer' signal and is required for envelope targeting.[108] The peptidase responsible for cleaving the C-terminal portion of the transit peptide has not been identified. Interestingly, a related Toc75 paralog that is also located at the outer envelope membrane of chloroplasts, OEP80, is not targeted by a cleavable transit peptide, nor does it appear to use the Toc-Tic apparatus. Rather, it uses intrinsic targeting information and appears to require proteinaceous components other than those of the Toc complex for its targeting.[109]

Targeting to the Intermembrane Space Utilizes a Unique Pathway

Tic22 is the only protein of the intermembrane space whose targeting has been studied. In fact, Tic22 is the only protein that has been conclusively shown to be a resident of the intermembrane space of plastids. Using a series of chimeric proteins, it was shown that the cleavable N-terminal presequence of Tic22 is both necessary and sufficient to direct the protein to the intermembrane space, and that the import of Tic22 is an energy-dependent process.[110] However, a stromal protein that is known to use the Toc-Tic import pathway did not compete for import of Tic22 into isolated chloroplasts, indicating that Tic22 uses a novel, previously unidentified pathway (Table 1). The components of this pathway have not been identified, and it remains to be seen if all intermembrane space proteins use the same pathway.

The Import of Specialized Proteins May Be Regulated

The levels of most nuclear-encoded plastid proteins are tightly regulated at the level of transcription and/or translation. Therefore, there is not apparent need for regulation of the protein import process. However, there are two exceptions to this generality. NADPH-dependent protochlorophyllide oxidoreductase (POR) is a critical enzyme in chlorophyll biosynthesis, converting its substrate protochlorophyllide (Pchilde) to chlorophyllide in a light-dependent reaction.[111] One isoform of this enzyme, PORA, plays a central role in plant development by controlling the availability of chlorophyll for the assembly of the photosynthetic apparatus during photomorphogenesis. Several reports indicate that the import of PORA is regulated in response to the availability of its substrate, Pchlide, within plastids of certain tissues.[112-114] The strict coordination of PORA import and Pchlide accumulation may be necessary to prevent the accumulation of chlorophyll intermediates that can result in photo-oxidative damage of the organelle. There are conflicting reports as to whether or not PORA import relies on the Toc-Tic machinery or utilizes a unique import system.[115-117] A complex in barley denoted the Pchlide-dependent translocon complex (Ptc) that includes the outer membrane proteins OEP16 and a Toc34-related protein has been implicated in PORA import (Table 1).[118,119]

It also has been proposed that putative Tic complex components Tic62, Tic55 and now Tic32/IEP32 are involved in sensing the redox state of the chloroplast, and in turn regulating the translocation of certain preproteins.[73] The import of members of at least two families of redox proteins (ferredoxin and ferredoxin-NAD(P)+ oxidoreductase) appear to be regulated in response to conditions that alter the redox state of plastids.[71,120] More work will be required to conclusively show that these putative Tic proteins are involved directly in the mechanism and/or regulation of import of these proteins.

Some Preproteins Destined for the Plastid Interior Might Be Synthesized without Transit Peptides

Until recently, all nuclear-encoded proteins destined for the interior compartments of plastids were thought to contain cleavable N-terminal transit peptides. However, a proteomic analysis of chloroplast envelopes and the power of bioinformatics recently led to the identification of ceQORH (chloroplast envelope Quinone OxidoReductase Homologue), a protein associated with the inner envelope membrane that appears to be targeted to chloroplasts without the use of a 'classic' transit peptide (Table 1).[121] This is the first example of a chloroplast protein targeted to an internal chloroplast subcompartment other than the outer envelope membrane that does not contain a cleavable transit peptide. More recently it has been reported that IEP32/Tic32, also an inner membrane protein, is targeted to chloroplasts without the use of a cleavable transit peptide and independently of the Toc-Tic pathway.[74] It has been postulated that IEP32/Tic32 is inserted into the inner membrane from the intermembrane space, whereas ceQORH is thought to first enter the stroma before being inserted into the membrane. The exact nature of the targeting signals remains to be deciphered. It has been shown that a portion of the N-terminal is required in both cases, but in the case of ceQORH is not sufficient for directing the protein to the chloroplast. Further characterization of these pathways and identification of other substrates will be of interest in the future.

Multiple Import Pathways Are Essential for Plastid Biogenesis during Plant Development

In vitro biochemical studies using chloroplasts isolated from seedlings of the model plant *Pisum sativum* (garden pea) led to the "general import pathway" hypothesis, which postulated that there existed one complex responsible for recognizing and translocating all classes of preproteins, at all stages of development, into all types of plastid.[19] With the advent of the model plant *Arabidopsis thaliana*, the use of molecular genetics and the availability of knockout mutants, this hypothesis has recently been challenged, and now seems unlikely. Indeed, the discovery of distinct import pathways and structurally distinct Toc complexes with preferences for specific classes of preproteins in chloroplasts of Arabidopsis seedlings has led to the suggestion that the import apparatus plays a direct role in the biogenesis and differentiation of plastids.[38] Therefore, a much more dynamic picture of the import apparatus is beginning to emerge.

Multiple isoforms of the Toc GTPases that are regulated differentially in response to specific developmental conditions have recently been discovered. The Toc159 family consists of four genes in Arabidopsis: *atTOC90, atTOC120, atTOC132,* and *atTOC159.* [37,122] atToc159 and atToc120/132 represent functionally distinct preprotein targeting pathways.[38,123,124] These receptors differentially associate with the two Arabidopsis Toc34 isoforms, atToc33 and atToc34, to generate structurally-distinct Toc translocons.[38]

In vivo analysis of Arabidopsis null mutants for each of the Toc159 family members indicates that atToc159 is specifically required for the import of photosynthesis-related proteins during chloroplast biogenesis.[37,38,123] In contrast, atToc120 and atToc132 appear to be required for the import of essential constitutively-expressed plastid proteins in all tissues.[38,123] This hypothesis is supported by the fact that the receptors exhibit distinct specificities for different classes of preproteins in accordance with their roles in plastid biogenesis.[23,38] The role of atToc90 is unclear because null mutants do not exhibit obvious phenotypes even in combination with null mutants of other Toc159 family members.[123,125] Although atToc33 and atToc34 can substitute for one another in vivo, they also appear to exhibit selectivity in binding to different preproteins.[27,126] These data suggest that the combinations of Toc GTPases contribute to the characteristic binding specificities of individual Toc complexes.

The presence of structurally and functionally distinct plastid protein targeting pathways likely reflects the need to maintain balanced import of a diverse array of preproteins and the dramatic changes in substrate levels that occur during plastid development. In addition, these distinct pathways may be specialized to provide a level of regulation for the import of specific subsets of preproteins, a function that may be critical for the maintenance of basic plastid function regardless of the developmental state of the organelle.

Conclusion

Components of the Toc and Tic translocon complexes that are responsible for the import of nuclear-encoded plastid proteins were first identified in the early-mid-1990s.[30,127] In the years since, many more components have been identified, and the molecular functions of many have been determined. However, partially due to the emergence of Arabidopsis as a model system, the sequencing of its genome and the availability of knockout mutants, the field remains one of intense and exciting investigation. Indeed, new putative Toc and Tic components continue to be identified, as do new pathways for targeting proteins to many chloroplast subcompartments. Furthermore, the discovery of distinct and specialized Toc complexes that are involved in the recognition and import of discrete sets of preproteins and may be involved in regulating the import of these proteins as part of the differentiation programs of different plastid types ensures that many more exciting discoveries in the field are still to come.

Acknowledgements

This work was supported by National Institutes of Health grant GM61893 and National Science Foundation grant MCB-0090727 to D.J.S.

References

1. Thompson WW, Whatley JM. Development of nongreen plastids. Annu Rev Plant Physiol 1980; 31:375-394.
2. Kirk JTO, Tilney-Bassett RAE. The Plastids: Their chemistry, structure, growth and inheritance. New York: Elsevier/North-Holland Biomedical Press, 1978.
3. Dyall SD, Brown MT, Johnson PJ. Ancient invasions: From endosymbionts to organelles. Science 2004; 304:253-257.
4. Sugiura M. The chloroplast genome. Plant Mol Biol 1992; 19:149-168.
5. Leister D. Chloroplast research in the genomic age. Trends Genet 2003; 19:47-56.
6. Kleffmann T, Russenberger D, von Zychlinski A et al. The Arabidopsis thaliana chloroplast proteome reveals pathway abundance and novel protein functions. Curr Biol 2004; 14:354-362.
7. Reumann S, Keegstra K. The endosymbiotic origin of the protein import machinery of chloroplastic envelope membranes. Trends Plant Sci 1999; 4:302-307.
8. Schnell DJ. Protein targeting to the thylakoid membrane. Annu Rev Plant Physiol Plant Mol Biol 1998; 49:97-126.
9. Mache R, Zhou D-X, S L-M et al. Nuclear control of early plastid differentiation. Plant Physiol Biochem 1997; 35:199-203.
10. Keegstra K, Cline K. Protein import and routing systems of chloroplasts. Plant Cell 1999; 11:557-570.
11. Bruce BD. The paradox of plastid transit peptides: Conservation of function despite divergence in primary structure. Biochim Biophys Acta 2001; 1541:2-21.
12. Emanuelsson O, Nielsen H, von Heijne G. ChloroP, a neural network-based method for predicting chloroplast transit peptides and their cleavage sites. Prot Sci 1999; 8:978-984.
13. Emanuelsson O, von Heijne G. Prediction of organellar targeting signals. Biochim Biophys Acta 2001; 1541:114-119.
14. Richly E, Leister D. An improved prediction of chloroplast proteins reveals diversities and commonalities in the chloroplast proteomes of Arabidopsis and rice. Gene 2004; 329:11-16.
15. Oblong JE, Lamppa GK. Identification of two structurally related proteins involved in proteolytic processing of precursors targeted to the chloroplast. EMBO J 1992; 11:4401-4409.
16. Richter S, Lamppa GK. Stromal processing peptidase binds transit peptides and initiates their ATP-dependent turnover in chloroplasts. J Cell Biol 1999; 147:33-43.
17. Chen X, Schnell DJ. Protein import into chloroplasts. Trends Cell Biol 1999; 9:222-227.
18. Keegstra K, Froehlich JE. Protein import into chloroplasts. Curr Opin Plant Biol 1999; 2:471-476.
19. Jarvis P, Soll J. Toc, tic, and chloroplast protein import. Biochim Biophys Acta 2002; 1590:177-189.

20. Schnell DJ, Blobel G, Keegstra K et al. A nomenclature for the protein import components of the chloroplast envelope. Trends Cell Biol 1997; 7:303-304.
21. Bauer J, Hiltbrunner A, Kessler F. Molecular biology of chloroplast biogenesis: Gene expression, protein import and intraorganellar sorting. Cell Mol Life Sci 2001; 58:420-433.
22. Schleiff E, Soll J, Kuchler M et al. Characterization of the translocon of the outer envelope of chloroplasts. J Cell Biol 2003; 160:541-551.
23. Smith MD, Rounds CM, Wang F et al. atToc159 is a selective transit peptide receptor for the import of nucleus-encoded chloroplast proteins. J Cell Biol 2004; 165:323-334.
24. Sveshnikova N, Soll J, Schleiff E. Toc34 is a preprotein receptor regulated by GTP and phosphorylation. Proc Natl Acad Sci USA 2000; 97:4973-4978.
25. Ma Y, Kouranov A, LaSala S et al. Two components of the chloroplast protein import apparatus, IAP86 and IAP75, interact with the transit sequence during the recognition and translocation of precursor proteins at the outer envelope. J Cell Biol 1996; 134:1-13.
26. Perry SE, Keegstra K. Envelope membrane proteins that interact with chloroplastic precursor proteins. Plant Cell 1994; 6:93-105.
27. Kubis S, Baldwin A, Patel R et al. The Arabidopsis ppi1 mutant is specifically defective in the expression, chloroplast import, and accumulation of photosynthetic proteins. Plant Cell 2003; 15:1859-1871.
28. Young ME, Keegstra K, Froehlich JE. GTP promotes the formation of early-import intermediates but is not required during the translocation step of protein import into chloroplasts. Plant Physiol 1999; 121:237-244.
29. Olsen LJ, Theg SM, Selman BR et al. ATP is required for the binding of precursor proteins to chloroplasts. J Biol Chem 1989; 264:6724-6729.
30. Schnell DJ, Kessler F, Blobel G. Isolation of components of the chloroplast protein import machinery. Science 1994; 266:1007-1012.
31. Hinnah SC, Hill K, Wagner R et al. Reconstitution of a chloroplast protein import channel. EMBO J 1997; 16:7351-7360.
32. Kouranov A, Schnell DJ. Analysis of the interactions of preproteins with the import machinery over the course of protein import into chloroplasts. J Cell Biol 1997; 139:1677-1685.
33. Sohrt K, Soll J. Toc64, a new component of the protein translocon of chloroplasts. J Cell Biol 2000; 148:1213-1221.
34. Chew O, Lister R, Qbadou S et al. A plant mitochondrial membrane protein with high amino acid sequence identity to a chloroplast protein import receptor. FEBS Lett 2004; 557:109-114.
35. Schleiff E, Jelic M, Soll J. A GTP-driven motor moves proteins across the outer envelope of chloroplasts. Proc Natl Acad Sci USA 2003; 100:4604-4609.
36. Leipe DD, Wolf YI, Koonin EV et al. Classification and evolution of p-loop GTPases and related ATPases. J Mol Biol 2002; 317:41-72.
37. Bauer J, Chen K, Hiltbunner A et al. The major protein import receptor of plastids is essential for chloroplast biogenesis. Nature 2000; 403:203-207.
38. Ivanova Y, Smith MD, Chen K et al. Members of the Toc159 import receptor family represent distinct pathways for protein targeting to plastids. Mol Biol Cell 2004; 15:3379-3392.
39. Constan D, Patel R, Keegstra K et al. An outer envelope membrane component of the plastid protein import apparatus plays an essential role in Arabidopsis. Plant J 2004; 38:93-106.
40. Kessler F, Schnell DJ. Chloroplast protein import: Solve the GTPase riddle for entry. Trends Cell Biol 2004; 14:334-338.
41. Sun YJ, Forouhar F, Li H-m et al. Crystal structure of pea Toc34, a novel GTPase of the chloroplast protein translocon. Nat Struct Biol 2002; 9:95-100.
42. Kessler F, Schnell DJ. A GTPase gate for protein import into chloroplasts. Nat Struct Biol 2002; 9:81-83.
43. Weibel P, Hiltbrunner A, Brand L et al. Dimerization of Toc-GTPases at the chloroplast protein import machinery. J Biol Chem 2003; 278:37321-37329.
44. Smith MD, Hiltbrunner A, Kessler F et al. The targeting of the atToc159 preprotein receptor to the chloroplast outer membrane is mediated by its GTPase domain and is regulated by GTP. J Cell Biol 2002; 159:833-843.

45. Becker T, Jelic M, Vojta A et al. Preprotein recognition by the Toc complex. EMBO J 2004; 23:520-530.
46. Olsen LJ, Keegstra K. The binding of precursor proteins to chloroplasts requires nucleoside triphosphates in the intermembrane space. J Biol Chem 1992; 267:433-439.
47. May T, Soll J. 14-3-3 proteins form a guidance complex with chloroplast precursor proteins in plants. Plant Cell 2000; 12:53-64.
48. Jackson-Constan D, Akita M, Keegstra K. Molecular chaperones involved in chloroplast protein import. Biochim Biophys Acta 2001; 1541:102-113.
49. Rial DV, Ottado J, Ceccarelli EA. Precursors with altered affinity for Hsp70 in their transit peptides are efficiently imported into chloroplasts. J Biol Chem 2003; 278:46473-46481.
50. Fulgosi H, Soll J. The chloroplast protein import receptors Toc34 and Toc159 are phosphorylated by distinct protein kinases. J Biol Chem 2002; 277:8934-8940.
51. Waegemann K, Soll J. Phosphorylation of the transit sequence of chloroplast precursor proteins. J Biol Chem 1996; 271:6545-6554.
52. Nakrieko KA, Mould RM, Smith AG. Fidelity of targeting to chloroplasts is not affected by removal of the phosphorylation site from the transit peptide. Eur J Biochem 2004; 271:509-516.
53. Hiltbrunner A, Bauer J, Vidi PA et al. Targeting of an abundant cytosolic form of the protein import receptor at Toc159 to the outer chloroplast membrane. J Cell Biol 2001; 154:309-316.
54. Bauer J, Hiltbrunner A, Weibel P et al. Essential role of the G-domain in targeting of the protein import receptor atToc159 to the chloroplast outer membrane. J Cell Biol 2002; 159:845-854.
55. Lee KH, Kim SJ, Lee YJ et al. The M domain of atToc159 plays an essential role in the import of proteins into chloroplasts and chloroplast biogenesis. J Biol Chem 2003; 278:36794-36805.
56. Reumann S, Davila-Aponte J, Keegstra K. The evolutionary origin of the protein-translocating channel of chloroplastic envelope membranes: Identification of a cyanobacterial homolog. Proc Natl Acad Sci USA 1999; 96:784-789.
57. Hinnah SC, Wagner R, Sveshnikova N et al. The chloroplast protein import cannel Toc75: Pore properties and interaction with transit peptides. Biophys J 2002; 83:899-911.
58. Kouranov A, Chen X, Fuks B et al. Tic20 and Tic22 are new components of the protein import apparatus at the chloroplast inner envelope membrane. J Cell Biol 1998; 143:991-1002.
59. Nielsen E, Akita M, Davila-Aponte J et al. Stable association of chloroplastic precursors with protein translocation complexes that contain proteins from both envelope membranes and a stromal Hsp100 molecular chaperone. EMBO J 1997; 16:935-946.
60. Scott SV, Theg SM. A new chloroplast proteins import intermediate reveals distinct translocation machineries in the two envelope membranes: Energetics and mechanistic implications. J Cell Biol 1996; 132:63-75.
61. Theg SM, Bauerle C, Olsen LJ et al. Internal ATP is the only energy requirement for the translocation of precursor proteins across chloroplastic membranes. J Biol Chem 1989; 264:6730-6736.
62. Pilon M, Schekman R. Protein translocation: How Hsp70 pulls it off. Cell 1999; 97:679-682.
63. Chen X, Smith MD, Fitzpatrick L et al. In vivo analysis of the role of atTic20 in protein import into chloroplasts. Plant Cell 2002; 14:641-654.
64. Heins L, Mehrle A, Hemmler R et al. The preprotein conducting channel at the inner envelope membrane of plastids. EMBO J 2002; 21:2616-2625.
65. Rassow J, Dekker PJ, van Wilpe S et al. The preprotein translocase of the mitochondrial inner membrane: Function and evolution. J Mol Biol 1999; 286:105-120.
66. Kessler F, Blobel G. Interaction of the protein import and folding machineries in the chloroplast. Proc Natl Acad Sci USA 1996; 93:7684-7689.
67. Inaba T, Li M, Alvarez-Huerta M et al. atTic110 functions as a scaffold for coordinating the stromal events of protein import into chloroplasts. J Biol Chem 2003; 278:38617-38627.
68. Akita M, Nielsen E, Keegstra K. Identification of protein transport complexes in the chloroplastic envelope membranes via chemical cross-linking. J Cell Biol 1997; 136:983-994.
69. Chou ML, Fitzpatrick LM, Tu SL et al. Tic40, a membrane-anchored cochaperone homolog in the chloroplast protein translocon. EMBO J 2003; 22:2970-2980.
70. Yalovsky S, Paulsen H, Michaeli D et al. Involvement of a chloroplast HSP70 heat shock protein in the integration of a protein (light-harvesting complex protein precursor) into the thylakoid membrane. Proc Natl Acad Sci USA 1992; 89:5616-5619.

71. Kuchler M, Decker S, Hormann F et al. Protein import into chloroplasts involves redox-regulated proteins. EMBO J 2002; 21:6136-6145.
72. Caliebe A, Grimm R, Kaiser G et al. The chloroplastic protein import machinery contains a Rieske-type iron-sulfur cluster and a monoculear iron-binding protein. EMBO J 1997; 16:7342-7350.
73. Hormann F, Kuchler M, Sveshnikov D et al. Tic32, an essential component in chloroplast biogenesis. J Biol Chem 2004; 279:34756-34762.
74. Nada A, Soll J. Inner envelope protein 32 is imported into chloroplasts by a novel pathway. J Cell Sci 2004; 117:3975-3982.
75. Cline K, Henry R. Import and routing of nucleus-encoded chloroplast proteins. Annu Rev Cell Dev Biol 1996; 12:1-26.
76. von Heijne G. Signal sequences: The limits of variation. J Mol Biol 1985; 184:99-105.
77. von Heijne G. A new method for predicting signal sequence cleavage sites. Nucl Acids Res 1986; 14:4683-4690.
78. Chaddock AM, Mant A, Karnauchov I et al. A new type of signal peptide: Central role of a twin-arginine motif in transfer signals for the delta-pH-dependent thylakoidal protein translocase. EMBO J 1995; 12:2715-2722.
79. Yuan J, Henry R, McCaffery M et al. SecA homolog in protein transport within chloroplasts: Evidence for endosymbiont-derived sorting. Science 1994; 266:796-798.
80. Laidler V, Chaddock AM, Knott RF et al. A SecY homolog in Arabadopsis thaliana. J Biol Chem 1995; 270:17664-17667.
81. Settles AM, Yonetani A, Baron A et al. Sec-Independent protein translocation by the maize Hcf106 protein. Science 1997; 278:1467-1470.
82. Mori H, Summer EJ, Ma X et al. Component specificity of the thylakoidal Sec and delta pH-dependent protein transport pathways. J Cell Biol 1999; 146:45-56.
83. Mori H, Summer EJ, Cline K. Chloroplast TatC plays a direct role in thylakoid (Delta) pH-dependent protein transport. FEBS Lett 2001; 501:65-68.
84. Hutcheon GW, Bolhuis A. The archaeal twin-arginine translocation pathway. Biochem Soc Trans 2003; 31:686-689.
85. Walter P, Johnson AE. Signal sequence recognition and protein targeting to the endoplasmic reticulum membrane. Annu Rev Cell Biol 1994; 10:87-119.
86. Tu C-J, Scheunemann D, Hoffmann NE. Chloroplast FtsY, chloroplast signal recognition particle, and GTP are required to reconstitute the soluble phase of light-harvesting chlorophyll protein transport into thylakoid membranes. J Biol Chem 1999; 274:27219-27224.
87. Kohorn BD, Tobin EM. A hydrphobic, carboxy-proximal region of a light-harvesting chlorophyll a/b protein is necessary for stable integration into thylakoid membranes. Plant Cell 1989; 1:159-166.
88. Sundberg E, Slagter JG, Fridborg I et al. ALBINO3, an Arabidopsis nuclear gene essential for chloropalst differentiation, encodes a chloroplast protein that shows homology to proteins present in bacterial membranes and yeast mitochondria. Plant Cell 1997; 9:717-730.
89. Moore M, Harrison MS, Peterson EC et al. Chloroplast Oxa1p homolog albino3 is required for post-translational integration of the light harvesting chlorophyll-binding protein into thylakoid membranes. J Biol Chem 2000; 275:1529-1532.
90. Nilsson R, van Wijk KJ. Transient interaction of cpSRP54 with elongating nascent chains of the chloroplast-encoded D1 protein; "cpSRP54 caught in the act". FEBS Lett 2002; 524:127-133.
91. Mant A, Woolhead CA, Moore M et al. Insertion of PsaK into the thylakoid membrane in a "Horseshoe" conformation occurs in the absence of signal recognition particle, nuceoside triphosphates, or functional albino3. J Biol Chem 2001; 276:36200-36206.
92. Robinson D, Karnauchov I, Hermann RG et al. Protease-sensitive thylakoidal import machinery for the Sec-, ΔpH-, and signal recognition particle-dependent protein import pathways, but not for CFoII integration. Plant J 1996;10:149-155.
93. Kim SJ, Robinson D, Robinson C. An Arabidopsis thaliana cDNA encoding PS II-X, a 4.1 kDa component of photosystem II: A bipartite presequence mediates SecA/delta pH-independent targeting into thylakoids. FEBS Letts 1996; 390:175-178.

94. Lorkovic ZJ, Schroder WP, Pakraski HB et al. Molecular characterization of PsbW, a nuclear-encoded component of the photosystem II reaction center complex in spinach. Proc Natl Acad Sci USA 1995; 92:8930-8934.
95. Lubeck J, Heins L, Soll J. A nuclear-encoded chloroplastic inner envelope membrane protein uses a soluble sorting intermediate upon import into the organelle. J Cell Biol 1997; 137:1279-1286.
96. Knight JS, Gray JC. The N-terminal hydrophobic region of the mature phosphate translocator is sufficient for targeting to the chloroplast inner envelope membrane. Plant Cell 1995; 7:1421-1432.
97. Pfanner N, Wiedemann N. Mitochondrial protein import: Two membranes, three translocases. Curr Opin Cell Biol 2002; 14:400-411.
98. Li H-m, Chen L-J. Protein targeting and integration signal for the chloroplastic outer envelope membrane. Plant Cell 1996; 8:2117-2126.
99. Li H-m, Moore T, Keegstra K. Targeting of proteins to the outer envelope membrane uses a different pathway than transport into chloroplasts. Plant Cell 1991; 3:709-717.
100. Tu S-H, Chen L-J, Smith MD et al. Import pathways of chloroplast interior proteins and outer-membrane proteins converge at Toc75. Plant Cell 2004; 16:2078-2088.
101. Chen D, Schnell DJ. Insertion of the 34-kDa chloroplast protein import component, IAP34, into the chloroplast outer membrane is dependent on its intrinsic GTP-binding capacity. J Biol Chem 1997; 272:6614-6620.
102. Froehlich JE, Benning C, Dormann P. The digalactosyldiacylglycerol (DGDG) synthase DGD1 is inserted into the outer envelope membrane of chloroplasts in a manner independent of the general import pathway and does not depend on direct interaction with monogalactosyldiacylglycerol synthase for DGDG biosynthesis. J Biol Chem 2001; 276:31806-31812.
103. Li H-m, Chen L-J. A novel chloroplastic outer membrane-targeting signal that functions at both termini of passenger polypeptides. J Biol Chem 1997; 272:10968-10974.
104. Qbadou S, Tien R, Soll J et al. Membrane insertion of the chloroplast outer envelope protein, Toc34: Constrains for insertion and topology. J Cell Sci 2003; 116:837-846.
105. Wallas TR, Smith MD, Sanchez-Nieto S et al. The roles of Toc34 and Toc75 in targeting the Toc159 preprotein receptor to chloroplasts. J Biol Chem 2003; 278:44289-44297.
106. Tranel PJ, Froehlich J, Goyal A et al. A component of the chloroplastic protein import apparatus is targeted to the outer envelope membrane via a novel pathway. EMBO J 1995; 14:2436-2446.
107. Tranel PJ, Keegstra K. A novel, bipartite transit peptide targets OEP75 to the outer membrane of the chloroplastic envelope. Plant Cell 1996; 8:2093-2104.
108. Inoue K, Keegstra K. A polyglycine stretch is necessary for proper targeting of the protein translocation channel precursor to the outer envelope membrane of chloroplasts. Plant J 2003; 34:661-669.
109. Inoue K, Potter D. The chloroplastic protein translocation channel Toc75 and its paralog OEP80 represent two distinct protein families and are targeted to the chloroplastic outer envelope by different mechanisms. Plant J 2004; 39:354-365.
110. Kouranov A, Wang H, Schnell DJ. Tic22 is targeted to the intermembrane space of chloroplasts by a novel pathway. J Biol Chem 1999; 274:25181-25186.
111. Beale SI. Enzymes of chlorophyll biosynthesis. Photosynth Res 1999; 60:43-73.
112. Reinbothe S, Runge S, Reinbothe C et al. Substrate-dependent transport of the NADPH:protochlorophyllide oxidoreductase into isolated plastids. Plant Cell 1995; 7:161-172.
113. Dahlin C, Aronsson H, Almkvist J et al. Protochlorophyllide-independent import of two NADPH: Pchilde oxidoreductase proteins (PORA and PORB) from barley into isolated plastids. Physiol Plant 2000; 109:298-303.
114. Kim C, Apel K. Substrate-dependent and organ-specific chloroplast protein import in planta. Plant Cell 2004; 16:88-98.
115. Reinbothe S, Mache R, Reinbothe C. A second, substrate-dependent site of protein import into chloroplasts. Proc Natl Acad Sci USA 2000; 97:9795-9800.
116. Aronsson H, Sohrt K, Soll J. NADPH:Protochlorophyllide oxidoreductase uses the general import route into chloroplasts. Biol Chem 2000; 381:1263-1267.
117. Aronsson H, Sundqvist C, Dahlin C. POR - import and membrane association of a key element in chloroplast development. Physiol Plant 2003; 118:1-9.

118. Reinbothe S, Quigley F, Gray J et al. Identification of plastid envelope proteins required for import of protochlorophyllide oxidoreductase A into the chloroplast of barley. Proc Natl Acad Sci USA 2004; 101:2197-2202.
119. Reinbothe S, Quigley F, Springer A et al. The outer plastid envelope protein Oep16: Role as precursor translocase in import of protochlorophyllide oxidoreductase A. Proc Natl Acad Sci USA 2004; 101:2203-2208.
120. Hirohashi T, Hase T, Nakai M. Maize nonphotosynthetic ferredoxin precursor is mis-targeted to the intermembrane space of chloroplasts in the presence of light. Plant Physiol 2001; 125:2154-2163.
121. Miras S, Salvi D, Ferro M et al. Noncanonical transit peptide for import into the chloroplast. J Biol Chem 2002; 277:47770-47778.
122. Hiltbrunner A, Bauer J, Alvarez-Huerta M et al. Protein translocon at the Arabidopsis outer chloroplast membrane. Biochem Cell Biol 2001; 79:629-635.
123. Kubis S, Patel R, Combe J et al. Functional specialization amongst the Arabidopsis Toc159 family of chloroplast protein import receptors. Plant Cell 2004; 16:2059-2077.
124. Vojta A, Alavi M, Becker T et al. The protein translocon of the plastid envelopes. J Biol Chem 2004; 279:21401-21405.
125. Hiltbrunner A, Grünig K, Alvarez-Huerta M et al. AtToc90, a new GTP-binding component of the Arabidopsis chloroplast protein import machinery. Plant Mol Biol 2004; 54:427-440.
126. Jelic M, Soll J, Schleiff E. Two Toc34 homologues with different properties. Biochemistry 2003; 42:5906-5916.
127. Kessler F, Blobel G, Patel HA et al. Identification of two GTP-binding proteins in the chloroplast protein import machinery. Science 1994; 266:1035-1039.

CHAPTER 9

The Mitochondrial Protein Import Machinery

Doron Rapaport*

Abstract

Mitochondria are surrounded by a double-membrane system that defines four intra-organelle compartments: the outer membrane, the inner membrane, the intermembrane space and the matrix. Hundreds of nuclear-encoded mitochondrial proteins are synthesized as precursor proteins in the cytosol and have to be targeted to and imported into the mitochondria. To facilitate this import process, precursor proteins contain targeting and sorting sequences which are recognized and decoded by mitochondrial translocation machineries. This chapter describes the mechanisms by which mitochondrial precursor proteins are targeted to the mitochondria, and sorted into the correct sub-mitochondrial compartment.

Introduction

Mitochondria are unique organelles which harbor numerous metabolic pathways and supply cells with energy in the form of ATP. Recently, it was established that mitochondria are pivotal in controlling cell life and death. Furthermore, over the past 10 years, mitochondrial defects have been implicated in a wide variety of degenerative diseases, aging, and cancer.[1]

Mitochondria are made up of the outer and inner membranes, which separate the intermembrane space (IMS) and the matrix from the cytosol. Mitochondria have been estimated to be composed of over 1000 different proteins in mammalian cells and of about 700-800 proteins in yeast.[2,3] Only 1-2% of these proteins are encoded by the mitochondrial genome and synthesized within the organelle itself. Therefore, importing precursor proteins into the organelle and sorting them into the correct sub-mitochondrial compartment are essential processes for mitochondrial biogenesis and, thereby, for eukaryotic cell viability (Fig. 1).

Most of our knowledge on the mitochondrial import machinery has been obtained studying the yeast *Saccharomyces cerevisiae* and the mold *Neurospora crassa,* but mitochondrial translocases from higher organisms such as mammals and plants contain components homologous to the fungal subunits, and appear to have a similar overall structure.[4-7]

Most of the mitochondrial precursor proteins, especially those destined for the matrix, are synthesized with an N-terminal extension, the presequence (also known as matrix targeting sequence). Presequences were shown to be sufficient to direct proteins to the mitochondria.[8] They are rich in positively-charged amino acid residues, ca. 15-50 residues long and have the potential to form amphiphilic α-helices. In contrast, all proteins of the mitochondrial outer membrane and some of the proteins destined to the inner membrane and the intermembrane

*Doron Rapaport—Institut für Physiologische Chemie der Universität München, München, Germany. Email: rapaport@bio.med.uni-muenchen.de

Protein Movement Across Membranes, edited by Jerry Eichler. ©2005 Eurekah.com and Springer Science+Business Media.

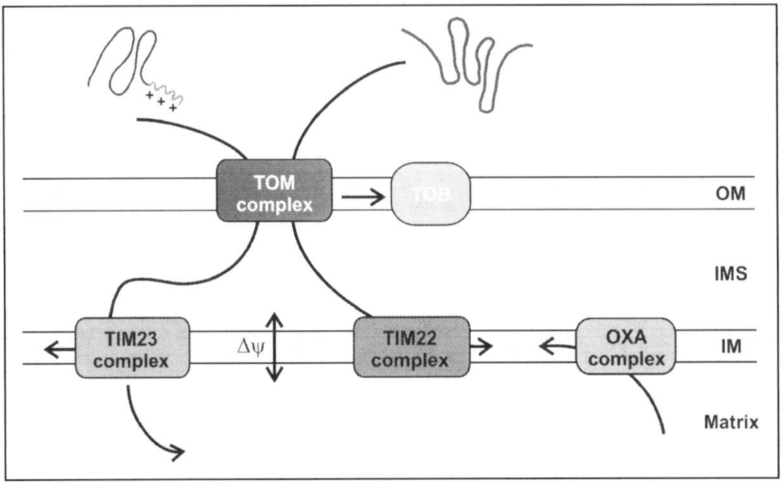

Figure 1. The mitochondrial protein import machinery. Mitochondrial precursor proteins interact initially with the translocase of the outer mitochondrial membrane (the TOM complex). Precursor proteins are further transferred to various translocation and sorting complexes. Beta-barrel precursors are transferred to the TOB complex in the outer membrane. Preproteins that contain a presequence are handed over to the TIM23 translocase at the inner membrane, whereas those inner membrane precursors that lack a presequence and contain multiple membrane-spanning segments are passed on to the TIM22 translocase in the inner membrane. Oxa1 is required for the insertion into the inner membrane of a number of mitochondrially-encoded proteins as well as some nuclear-encoded ones. OM, outer membrane; IMS, intermembrane space; IM, inner membrane.

space are devoid of a typical presequence. The targeting information in these proteins is rather contained in the protein sequence itself.

Co- versus Post-Translational Import

The question whether mitochondrial protein import occurs co- or post-translationally is still open. Supporting the cotranslational model is the observation that under normal growth conditions, fully synthesized, but unprocessed precursor proteins are essentially not detected in the cytosol, suggesting that they are translocated into mitochondria either very soon after synthesis or cotranslationally.[9] Furthermore, when translation is inhibited by addition of cycloheximide, yeast mitochondria are covered with ribosomes, suggesting that the ribosome-bound precursors are accumulated on the surface of mitochondria.[10] Thus, it seems that the relative kinetics of translation and translocation probably determines the enrichment of polysomes encoding mitochondrial precursors on the organelle surface.[11] On the other hand, convincing evidences for post-translational import also exist. First, many mitochondrial precursor proteins synthesized in a cell-free system can be imported post-translationally into isolated mitochondria. Second, mitochondrial precursor proteins, accumulated in yeast cells by dissipating the membrane potential ($\Delta\Psi$) across the mitochondrial inner membrane, can be subsequently chased into mitochondria by removal of the uncoupler.[12] Taken together, the translation of mitochondrial precursor proteins in the cytosol is generally not coupled to their import into the organelle, and the vast majority of precursor proteins can be imported post-translationally.

To ensure efficient import and to prevent aggregation of hydrophobic precursor proteins, cytosolic chaperones stabilize the precursors in an import-competent conformation.

These chaperones include Hsp70, Hsp40, nascent-associated polypeptide complex (NAC), ribosome-associated complex (RAC) and mitochondrial import stimulation factor (MSF).[7,11] Recently, a direct cooperation between cytosolic chaperones and the mitochondrial import machinery has been demonstrated to be part of the import pathway of precursors of metabolite carrier proteins.[13]

The Translocase of the Outer Membrane as the Gate to the Organelle

At the surface of mitochondria, precursor proteins are recognized by the translocase of the outer mitochondrial membrane (TOM complex) (Fig. 2). The TOM complex is involved in the import of all mitochondrial precursor proteins characterized so far. It has the capacity to insert some outer membrane proteins into or to translocate all other precursor proteins across the outer membrane. Import of preproteins into the inner membrane, matrix, and in some cases, the intermembrane space, requires the additional action of translocases of the inner mitochondrial membrane (TIM23, TIM22 and Oxa1 complexes).

Within the TOM complex, components with domains that are exposed to the cytosol function as preprotein receptors. Tom20 and Tom22 are involved in the translocation of most protein precursors, in particular those with presequences.[14,15] Another receptor, Tom70, forms a binding site for a more restricted set of precursor proteins, most notably the mitochondrial carrier family that is responsible for metabolite transport across the inner membrane.[16-18] The subunits Tom40, Tom22, Tom7, Tom6, and Tom5 are embedded in the outer membrane and form the TOM core complex, which is also called the general insertion pore (Fig. 2).[19-22] Tom40 interacts with polypeptide chains in transit and is the major component of the protein-conducting pore.[23-26] Tom22 and Tom5 serve dual functions; they link the initial receptors to the pore, thereby functioning as secondary receptors, and play a role in the integrity of the complex.[27-29] The two small components, Tom6 and Tom7, play a structural role in the organization of the TOM complex. Tom6 forms the link between Tom40 and Tom22, while Tom7 destabilizes association of the various subunits of the TOM complex.[30-32]

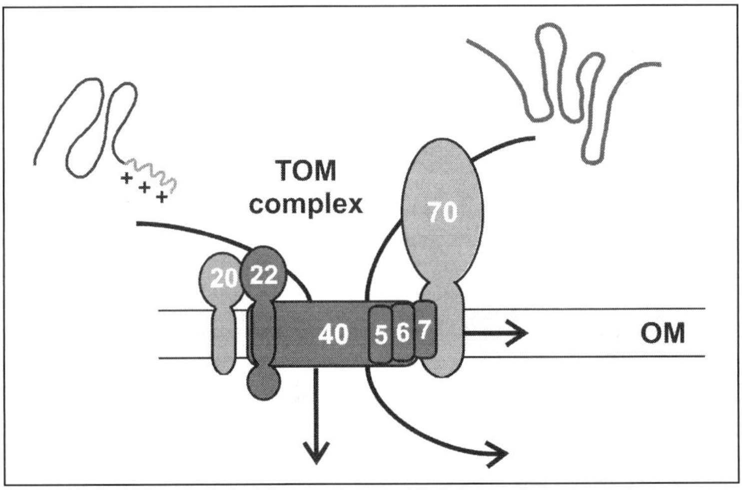

Figure 2. The translocase of the outer mitochondrial membrane (TOM complex). The TOM complex contains import receptors, Tom20 and Tom70. These two receptors are loosely attached to the rest of the complex (the TOM core complex). The TOM core complex builds the protein conducting pore and contains Tom5, Tom6, Tom7, Tom22 and Tom40. The names of the Tom subunits reflect their molecular weights.

Translocation of Preproteins Across the TOM Complex

Presequences interact initially with the primary receptor Tom20. The structure of a presequence in complex with the cytosolic domain of Tom20 suggests that the hydrophobic face of the helical presequence forms the contact with Tom20.[33] The cytosolic domain of Tom22 takes part in the formation of this surface binding site, termed the Cis site.[23,34,35] Crosslinking experiments suggest that the presequence is already in the vicinity of Tom40 at this early stage of import.[23,36] Movement of the presequence to the inner side of the outer membrane results in the formation of a second intermediate bound at the Trans site of the outer membrane. Unfolding of the translocating polypeptide chain, a process which is required for import into mitochondria, can occur at this stage.[36-39] Interestingly, in vitro binding experiments have demonstrated that the isolated TOM complex can transfer presequences into the translocation pore.[25] Thus, the TOM complex represents the minimal machinery for the recognition and partial translocation of precursor proteins. Both Tom40 and the C-terminal intermembrane space domain of Tom22 were suggested to bind the presequence at the Trans site.[23,36,40,41] Binding of the preprotein to the Cis or Trans sites induces distinct structural alterations in Tom40, and influences the interactions of Tom6 with both Tom40 and Tom22.[32,42]

What is the driving force for translocating presequence-containing preprotein across the outer membrane? The translocation of mitochondrial presequences across the inner membrane requires a membrane potential across the inner membrane and ATP hydrolysis by matrix chaperones. In contrast, there is no membrane potential across the outer membrane, and the membrane potential across the inner membrane is not required to translocate the presequence across the outer membrane.[37] Furthermore, translocation across the outer membrane does not depend on ATP hydrolysis. The current model suggests that a chain of presequence binding sites with increased affinity towards the presequence provide the driving force for translocation across the outer membrane and ensures vectorial movement of the precursor protein.

Insertion of Precursors of β-barrel Proteins into the Outer Membrane

A number of membrane-embedded β-barrel proteins made up from anti-parallel β-sheets constitute a distinct group of mitochondrial outer membrane proteins.[43,44] Other biological membranes that contain β-barrel proteins are the outer membranes of Gram-negative bacteria and of chloroplasts.[45-47] This most likely reflects the evolutionary origin of mitochondria from an endosymbiont that belonged to the class of Gram-negative bacteria.

The signals that target β-barrel precursor proteins to their sub-cellular location and the mechanism of their insertion into the membrane are only partially understood.[44,48] In the case of mitochondria, the precursors are initially recognized by the receptor components of the TOM complex (Tom20 and Tom70). They are then translocated through the import pore of the TOM complex (Fig. 3).[22,49-52] From the TOM complex, β-barrel precursors are transferred to the TOB/SAM complex.[53-55] Recent reports suggest that the small Tim proteins in the IMS are involved in this transfer process.[56,57] The major component of the TOB complex is Tob55 (also known as Sam50 and Omp85). Tob55 is essential for viability of yeast cells and promotes the insertion of β-barrel proteins into the mitochondrial outer membrane.[53,54,58] Tob55 has sequence similarity to the highly conserved bacterial protein Omp85 which was proposed to mediate the insertion of β-barrel proteins into the bacterial outer membrane.[59] Sequence analysis suggested that Tob55 homologues are found in the outer membrane of mitochondria of all eukaryotes.[54,58] Thus, the biogenesis of β-barrel proteins appears to be conserved from Bacteria to mammals. Electron microscopy and electrophysiological measurements suggest that Tob55 forms pore structures within the membrane.[54]

The other known components of the TOB complex are the outer membrane proteins Mas37 and Tob38/Sam35 (Fig. 3). Mas37 interacts with Tob55 and plays an as yet undefined

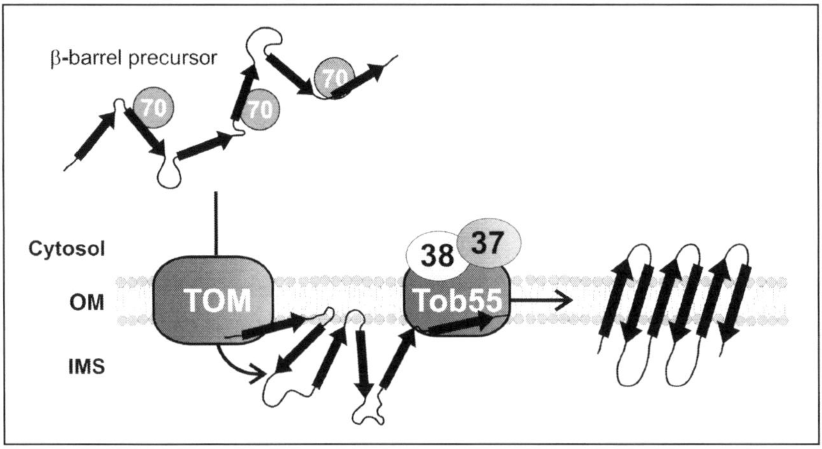

Figure 3. Insertion of β-barrel precursors into the mitochondrial outer membrane. Precursors of β-barrel proteins are kept in the cytosol in an import-competent conformation by cytosolic chaperones like Hsp70. They are initially recognized by the receptor components of the TOM complex before being translocated through the import pore of the TOM complex. From the TOM complex, β-barrel precursors are transferred to the TOB complex which mediates insertion into the outer membrane and consists of Tob55, Tob38 and Mas37. The major component, Tob55, forms probably pore structures within the membrane.

role in the biogenesis of β-barrel proteins.[53-55] Tob38 is essential for viability in yeast and crucial for the biogenesis of mitochondrial β-barrel proteins.[60,61] Together with Tob55, Tob38 forms a functional TOB core complex and is essential for the integrity and function of the TOB complex.[60]

Translocation of Presequence-Containing Preproteins Across the Inner Membrane

Precursor proteins harboring a presequence are transferred from the TOM complex to the TIM23 machinery (Fig. 4). This machinery is built from a membrane-embedded part and associated motor. The membrane-integrated portion is composed of three essential proteins: Tim50, Tim17 and Tim23. Tim50 spans the inner membrane once with the majority of the protein found in the IMS. As the presequence emerges from the TOM complex, it interacts with Tim50, which mediates presequence transfer to the pore-forming component, Tim23.[62-64] Tim23 forms a dimer through its N-terminal domain which also binds presequences.[65] The membrane potential across the inner membrane activates the channel formed by Tim23 and exerts an electrophoretic force on the positively-charged presequence, thereby driving it across the inner membrane.[66,67] Tim17 is tightly associated with Tim23 and may regulate the channel activity of the latter. Its precise function, however, is not yet resolved.

Some presequence-containing preproteins are destined to the inner membrane or the IMS. These proteins possess a hydrophobic membrane-spanning segment downstream of their presequence. Whereas the presequence portion of such proteins could be translocated into the matrix, the hydrophobic sequence arrests as a stop-transfer signal, halting the subsequent translocation of the rest of the protein molecule. These proteins are then released into the lipid bilayer where they stay as integral proteins. In the case of some proteins like cytochrome b_2, Mgm1 and Mcr1, the membrane-integrated intermediate is further processed by inner membrane-embedded peptidases which cleave off the transmembrane domain, thereby releasing a soluble domain into the IMS.[68-71]

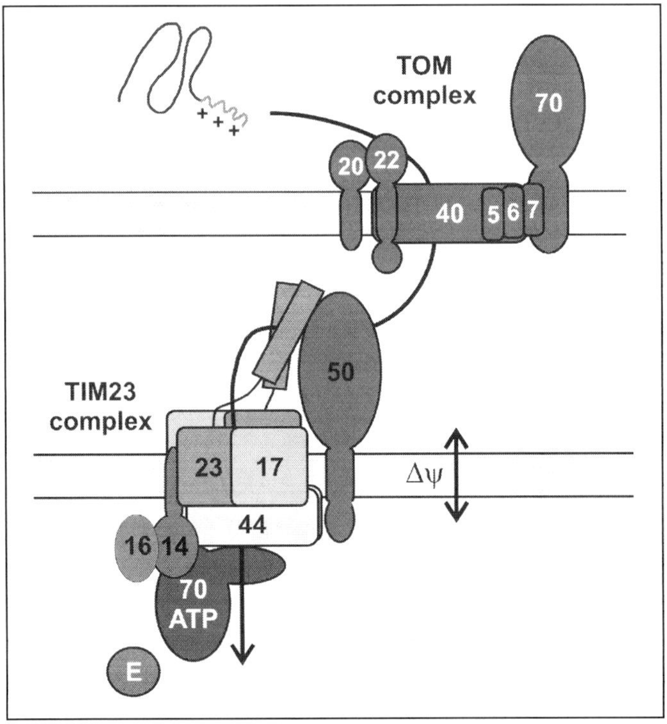

Figure 4. Import of preproteins by the TIM23 pathway. The presequence is first recognized by the receptors of the TOM complex before being translocated through the import pore of the complex. As the preprotein emerges from the TOM complex, it interacts with Tim50 which mediates transfer of the precursor to the Tim23-Tim17 channel. MtHsp70 and the other components of the import motor ensure the complete translocation of the precursor protein into the matrix.

The majority of presequence-containing preproteins are imported into the matrix. This import is mediated by both parts of the TIM23 machinery; the membrane-embedded portion and the ATP-driven motor apparatus associated with the former part at the matrix side of the inner membrane (Fig. 4).[72,73] The motor is composed of five proteins, all of them essential for viability of yeast cells. Tim44 makes contact with Tim23 and recruits three other components of the import motor to the translocation channel, mtHsp70 and the cochaperones Tim14 and Tim16 (also known as Pam18 and Pam16, respectively). The chaperone mtHsp70 binds to a polypeptide as it exits from the translocation channel. The vectorial movement of the polypeptide chain across the inner membrane is achieved by repeated ATP-driven cycles of binding and release of mtHsp70 to Tim44 and to the polypeptide chain.[72-74] For this function, Tim44 and mtHsp70 are dependent on three additional cochaperones, the nucleotide exchange factor Mge1, the J domain-related protein Tim16 and the J domain-related Tim16.[75-78] Tim14 is anchored to the inner membrane by a single transmembrane segment and exposes a J domain to the matrix side. It interacts with Tim44 and mtHsp70 in an ATP-dependent manner and stimulates the ATPase activity of mtHsp70. Thus, it allows rapid and efficient trapping of the precursor protein by mtHsp70. Tim16 is required for the integrity of the import motor.

Once the presequence is fully exposed to the matrix, it is cleaved by the mitochondrial processing peptidase (MPP), composed of two essential subunits.[79]

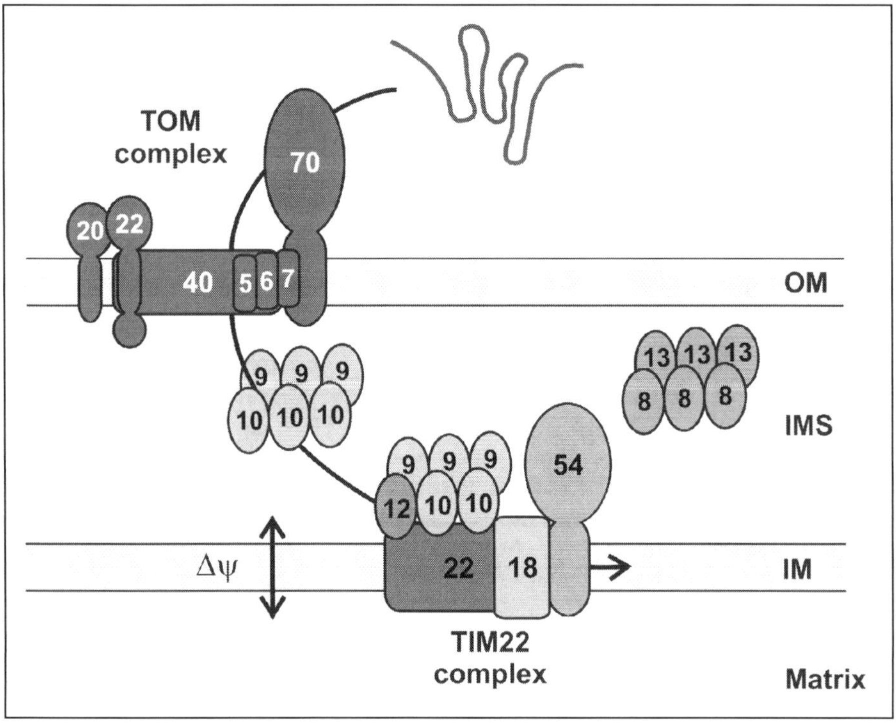

Figure 5. Import of polytopic proteins by the TIM22 pathway. Polytopic inner membrane proteins contain internal import signals and are recognized initially by the Tom70 receptor. Their translocation across the outer membrane is assisted by the Tim9/Tim10 complex which guides them to the TIM22 complex. The TIM22 translocase requires the membrane potential across the inner membrane to mediate the insertion of polytopic precursor proteins into the inner membrane.

Insertion of Polytopic Proteins into the Inner Membrane

The TIM22 import pathway mediates the import and insertion into the inner membrane of polytopic proteins like those of the mitochondrial carrier family and the import components Tim17, Tim22 and Tim23 (Fig. 5).[80,81] After their synthesis on cytosolic ribosomes, these precursor proteins are protected against aggregation by interaction with molecular chaperones which also help to direct them to the import receptor Tom70.[13,82] From the receptor protein, the carrier precursor is transferred to the outer membrane translocation pore. Carrier precursor proteins, like other mitochondrial precursors with internal import signals, cross the TOM complex in a loop structure.[18,83-85]

Upon their exit from the TOM complex, these precursors interact with complexes of the 'small Tim' components (Fig. 5). Tim9 and Tim10 are two such proteins. These form a heterohexamer of about 70 kDa which contains three Tim9 and three Tim10 molecules.[86-88] Two other small Tim components, Tim8 and Tim13, also form a heterohexameric complex in the intermembrane space.[89,90] In contrast to Tim9/Tim10, Tim8/Tim13 are not essential for viability, and seem to be involved in the import of only a limited number of precursor proteins.[84,91-93] The small Tim proteins were proposed to serve as chaperones that prevent the aggregation of the hydrophobic precursors in the IMS. They facilitate the transfer of precursors across the IMS and deliver them to the membrane-embedded TIM22 complex.[94-98]

The TIM22 complex contains three integrated membrane proteins: Tim22, Tim18 and Tim54. Tim22 is an essential protein, which has sequence similarity to Tim17 and Tim23. It is the core component of the TIM22 complex and was demonstrated to form pores in lipid bilayers with two distinct conducting states.[99] The larger state may represent a conformation where two α-helices could be accommodated within the channel. Thus, it may be that substrates of Tim22 are inserted in a loop structure into the channel before being released into the lipid bilayer. The membrane potential across the inner membrane is the only external energy required for membrane insertion by the TIM22 complex. In contrast to the TIM23 complex, no requirement for ATP hydrolysis was reported. Finally, Tim18 and Tim54 may stabilize the oligomeric structure of the TIM22 complex.

The Oxa1 Machinery

The Oxa1 protein belongs to an evolutionary conserved group of proteins called the Alb3/Oxa1/YidC family that perform an important function in catalyzing the insertion and assembly of membrane proteins in diverse biological systems like Bacteria and chloroplasts.[100] Oxa1 mediates, by an unknown mechanism, the insertion into the mitochondrial inner membrane of mitochondrially- and nuclear-encoded proteins which are exported from the matrix.[101,102] Nucleus-encoded proteins that use the Oxa1 pathway are initially completely imported into the mitochondrial matrix by the TOM-TIM23 pathway. They are then integrated into the inner membrane by a process which depends on Oxa1. The Oxa1-mediated insertion of mitochondrially-encoded proteins was suggested to be a cotranslational process. This insertion is facilitated by a direct interaction of the C-terminal domain of Oxa1 with mitochondrial ribosomes.[103,104]

Concluding Remarks

Mitochondria contain a complex machinery for import and sorting of mitochondrial precursor proteins. All these precursors are initially recognized by the TOM complex at the outer membrane and are then further translocated by additional, multi-subunit sorting and import complexes until they reach their residential sub-mitochondrial compartment. The importance of these processes is demonstrated by the fact that many components of the involved translocases are essential for the viability of the eukaryotic cell.

Despite the fast progress in the field in the last decade, many questions remain still unanswered. Characterization of the internal targeting and sorting signals and understanding how are they are recognized by the mitochondrial machinery are just two of many such unresolved topics. Future studies in the field can address more clinically-oriented subjects, like potential involvement of import components in programmed cell death or dysfunction of such components as potential causes of mitochondrial diseases.

References

1. Wallace DC. Mitochondrial diseases in man and mouse. Science 1999; 283:1482-1488.
2. Sickmann A, Reinders J, Wagner Y et al. The proteome of Saccharomyces cerevisiae mitochondria. Proc Natl Acad Sci USA 2003; 100:13207-13212.
3. Taylor SW, Fahy E, Zhang B et al. Characterization of the human heart mitochondrial proteome. Nat Biotechnol 2003; 21:281-286.
4. Mori M, Terada K. Mitochondrial protein import in animals. Biochim Biophys Acta 1998; 1403:12-27.
5. Bauer MF, Rothbauer U, Muhlenbein N et al. The mitochondrial TIM22 preprotein translocase is highly conserved throughout the eukaryotic kingdom. FEBS Lett 1999; 464:41-47.
6. Braun HP, Schmitz UK. The protein-import apparatus of plant mitochondria. Planta 1999; 209:267-274.

7. Hoogenraad NJ, Ward LA, Ryan MT. Import and assembly of proteins into mitochondria of mammalian cells. Biochim Biophys Acta 2002; 1592:97-105.
8. Neupert W. Protein import into mitochondria. Annu Rev Biochem 1997; 66:863-917.
9. Fujiki M, Verner K. Coupling of cytosolic protein synthesis and mitochondrial protein import in yeast: Evidence for cotranslational import in vivo. J Biol Chem 1993; 268:1914-1920.
10. Kellems RE, Allison VF, Butow RA. Cytoplasmic type 80S ribosomes associated with yeast mitochondria. J Cell Biol 1975; 65:1-14.
11. Beddoe T, Lithgow T. Delivery of nascent polypeptides to the mitochondrial surface. Biochim Biophys Acta 2002; 1592:35-39.
12. Reid GA, Schatz G. Import of proteins into mitochondria. Yeast cells grown in the presence of carbonyl cyanide-m-chlorophenyl-hydrazone accumulate massive amounts of some mitochondrial precursor polypeptides. J Biol Chem 1982; 257:13056-13061.
13. Young JC, Hoogenraad NJ, Hartl F-U. Molecular chaperones Hsp90 and Hsp70 deliver preproteins to the mitochondrial import receptor Tom70. Cell 2003; 112:41-50.
14. Harkness TA, Nargang FE, van der Klei I et al. A crucial role of the mitochondrial protein import receptor MOM19 for the biogenesis of mitochondria. J Cell Biol 1994; 124:637-648.
15. Lithgow T, Glick BS, Schatz G. The protein import receptor of mitochondria. Trend Biochem Sci 1995; 20:98-101.
16. Schlossmann J, Dietmeier K, Pfanner N et al. Specific recognition of mitochondrial preproteins by the cytosolic domain of the import receptor MOM72. J Biol Chem 1994; 269:11893-11901.
17. Brix J, Rudiger S, Bukau B et al. Distribution of binding sequences for the mitochondrial import receptors Tom20, Tom22, and Tom70 in a presequence-carrying preprotein and a noncleavable preprotein. J Biol Chem 1999; 274:16522-16530.
18. Wiedemann N, Pfanner N, Ryan MT. The three modules of ADP/ATP carrier cooperate in receptor recruitment and translocation into mitochondria. EMBO J 2001; 20:951-960.
19. Dekker PJT, Ryan MT, Brix J et al. Preprotein translocase of the outer mitochondrial membrane: Molecular dissection and assembly of the general import pore complex. Mol Cell Biol 1998; 18:6515-6524.
20. Künkele K-P, Heins S, Dembowski M et al. The preprotein translocation channel of the outer membrane of mitochondria. Cell 1998; 93:1009-1019.
21. Ahting U, Thun C, Hegerl R et al. The TOM core complex: The general protein import pore of the outer membrane of mitochondria. J Cell Biol 1999; 147:959-968.
22. Rapaport D. Biogenesis of the mitochondrial TOM complex. Trends Biochem Sci 2002; 26:191-197.
23. Rapaport D, Neupert W, Lill R. Mitochondrial protein import. Tom40 plays a major role in targeting and translocation of preproteins by forming a specific binding site for the presequence. J Biol Chem 1997; 272:18725-18731.
24. Hill K, Model K, Ryan MT et al. Tom40 forms the hydrophilic channel of the mitochondrial import pore for preproteins. Nature 1998; 395:516-521.
25. Stan T, Ahting U, Dembowski M et al. Recognition of preproteins by the isolated TOM complex of mitochondria. EMBO J 2000; 19:4895-4902.
26. Ahting U, Thieffry M, Engelhardt H et al. Tom40, the pore-forming component of the protein-conducting TOM channel in the outer membrane of mitochondria. J Cell Biol 2001; 153:1151-1160.
27. Dietmeier K, Hönlinger A, Bömer U et al. Tom5 functionally links mitochondrial preprotein receptors to the general import pore. Nature 1997; 388:195-200.
28. van Wilpe S, Ryan MT, Hill K et al. Tom22 is a multifunctional organizer of the mitochondrial preprotein translocase. Nature 1999; 401:485-489.
29. Habib SJ, Vasiljev A, Neupert W et al. Multiple functions of tail-anchor domains of mitochondrial outer membrane proteins. FEBS Lett 2003; 555:511-515.
30. Alconada A, Kübrich M, Moczko M et al. The mitochondrial receptor complex: The small subunit Mom8b/Isp6 supports association of receptors with the general insertion pore and transfer of preproteins. Mol Cell Biol 1995; 15:6196-6205.
31. Hönlinger A, Bömer U, Alconada A et al. Tom7 modulates the dynamics of the mitochondrial outer membrane translocase and plays a pathway-related role in protein import. EMBO J 1996; 15:2125-2137.

32. Dembowski M, Künkele K-P, Nargang FE et al. Assembly of Tom6 and Tom7 into the TOM core complex of Neurospora crassa. J Biol Chem 2001; 276:17679-17685.
33. Abe Y, Shodai T, Muto T et al. Structural basis of presequence recognition by the mitochondrial protein import receptor Tom20. Cell 2000; 100:551-560.
34. Hönlinger A, Kübrich M, Moczko M et al. The mitochondrial receptor complex: Mom22 is essential for cell viability and directly interacts with preproteins. Mol Cell Biol 1995; 15:3382-3389.
35. Mayer A, Nargang FE, Neupert W et al. MOM22 is a receptor for mitochondrial targeting sequences and cooperates with MOM19. EMBO J 1995; 14:4204-4211.
36. Kanamori T, Nishikawa S-I, Nakai M et al. Uncoupling of transfer of the presequence and unfolding of the mature domain in precursor translocation across the mitochondrial outer membrane. Proc Natl Acad Sci USA 1999; 96:3634-3639.
37. Mayer A, Neupert W, Lill R. Mitochondrial protein import: Reversible binding of the presequence at the trans side of the outer membrane drives partial translocation and unfolding. Cell 1995; 80:127-137.
38. Rapaport D, Mayer A, Neupert W et al. Cis and trans sites of the TOM complex in unfolding and initial translocation of preproteins. J Biol Chem 1998; 273:8806-8813.
39. Esaki M, Kanamori T, Nishikawa S et al. Tom40 protein import channel binds to nonnative proteins and prevents their aggregation. Nat Struct Biol 2003; 10:988-994.
40. Bolliger L, Junne T, Schatz G et al. Acidic receptor domains on both sides of the outer membrane mediate translocation of precursor proteins into yeast mitochondria. EMBO J 1995; 14:6318-6326.
41. Moczko M, Bömer U, Kübrich M et al. The intermembrane space domain of mitochondrial Tom22 functions as a trans binding site for preproteins with N-terminal targeting sequences. Mol Cell Biol 1997; 17:6574-6584.
42. Rapaport D, Künkele K-P, Dembowski M et al. Dynamics of the TOM complex of mitochondria during binding and translocation of preproteins. Mol Cell Biol 1998; 18:5256-5262.
43. Gabriel K, Buchanan SK, Lithgow T. The alpha and beta: Protein translocation across mitochondrial and plastid outer membranes. Trends Biochem Sci 2001; 26:36-40.
44. Rapaport D. How to find the right organelle-targeting signals in mitochondrial outer membrane proteins. EMBO Rep 2003; 4:948-952.
45. Tamm LK, Arora A, Kleinschmidt JH. Structure and assembly of β-barrel membrane proteins. J Biol Chem 2001; 276:32399-32402.
46. Schleiff E, Eichacker LA, Eckart K et al. Prediction of the plant beta-barrel proteome: A case study of the chloroplast outer envelope. Protein Sci 2003; 12:748-759.
47. Wimley WC. The versatile β-barrel membrane protein. Curr Opin Struct Biol 2003; 13:404-411.
48. Johnson AE, Jensen RE. Barreling through the membrane. Nat Struct Mol Biol 2004; 11:113-114.
49. Rapaport D, Neupert W. Biogenesis of Tom40, core component of the TOM complex of mitochondria. J Cell Biol 1999; 146:321-331.
50. Krimmer T, Rapaport D, Ryan MT et al. Biogenesis of the major mitochondrial outer membrane protein porin involves a complex import pathway via receptors and the general import pore. J Cell Biol 2001; 152:289-300.
51. Model K, Meisinger C, Prinz T et al. Multistep assembly of the protein import channel of the mitochondrial outer membrane. Nat Struct Biol 2001; 8:361-370.
52. Schleiff E, Silvius JR, Shore GC. Direct membrane insertion of voltage-dependent anion-selective channel protein catalyzed by mitochondrial Tom20. J Cell Biol 1999; 145:973-978.
53. Kozjak V, Wiedemann N, Milenkovic D et al. An essential role of Sam50 in the protein sorting and assembly machinery of the mitochondrial outer membrane. J Biol Chem 2003; 278:48520-48523.
54. Paschen SA, Waizenegger T, Stan T et al. Evolutionary conservation of biogenesis of β-barrel membrane proteins. Nature 2003; 426:862-866.
55. Wiedemann N, Kozjak V, Chacinska A et al. Machinery for protein sorting and assembly in the mitochondrial outer membrane. Nature 2003; 424:565-571.
56. Wiedemann N, Truscott KN, Pfannschmidt S et al. Biogenesis of the protein import channel Tom40 of the mitochondrial outer membrane: Intermembrane space components are involved in an early stage of the assembly pathway. J Biol Chem 2004; 279:18188-18194.
57. Hoppins SC, Nargang FE. The Tim8-Tim13 complex of Neurospora crassa functions in the assembly of proteins into both mitochondrial membranes. J Biol Chem 2004; 279:12396-12405.

58. Gentle I, Kipros G, Beech P et al. The Omp85 family of proteins is essential for outer membrane biogenesis in mitochondria and bacteria. J Cell Biol 2004; 164:19-24.
59. Voulhoux R, Bos MP, Geurtsen J et al. Role of a highly conserved bacterial protein in outer membrane protein assembly. Science 2003; 299:262-265.
60. Waizenegger T, Habib SJ, Lech M et al. Tob38, a novel essential component in the biogenesis of β-barrel proteins of mitochondria. EMBO Rep 2004; 7:704-709.
61. Milenkovic D, Kozjak V, Wiedemann N et al. Sam35 of the mitochondrial protein sorting and assembly machinery is a peripheral outer membrane protein essential for cell viability. J Biol Chem 2004; 279:22781-22785.
62. Yamamoto H, Esaki M, Kanamori T et al. Tim50 is a subunit of the TIM23 complex that links protein translocation across the outer and inner mitochondrial membranes. Cell 2002; 111:519-528.
63. Geissler A, Chacinska A, Truscott KN et al. The mitochondrial presequence translocase: An essential role of Tim50 in directing preproteins to the import channel. Cell 2002; 111:507-518.
64. Mokranjac D, Paschen SA, Kozany C et al. Tim50, a novel component of the TIM23 preprotein translocase of mitochondria. EMBO J 2003; 22:816-825.
65. Bauer MF, Sirrenberg C, Neupert W et al. Role of Tim23 as voltage sensor and presequence receptor in protein import into mitochondria. Cell 1996; 87:33-41.
66. Truscott KN, Kovermann P, Geissler A et al. A presequence- and voltage-sensitive channel of the mitochondrial preprotein translocase formed by Tim23. Nat Struct Biol 2001; 8:1074-1082.
67. Huang S, Ratliff KS, Matouschek A. Protein unfolding by the mitochondrial membrane potential. Nat Struct Biol 2002; 9:301-307.
68. Glick BS, Brandt A, Cunningham K et al. Cytochromes c1 and b2 are sorted to the intermembrane space of yeast mitochondria by a stop-transfer mechanism. Cell 1992; 69:809-822.
69. Hahne K, Haucke V, Ramage L et al. Incomplete arrest in the outer membrane sorts NADH-cytochrome b5 reductase to two different submitochondrial compartments. Cell 1994; 79:829-839.
70. McQuibban GA, Saurya S, Freeman M. Mitochondrial membrane remodelling regulated by a conserved rhomboid protease. Nature 2003; 423:537-541.
71. Herlan M, Bornhovd C, Hell K et al. Alternative topogenesis of Mgm1 and mitochondrial morphology depend on ATP and a functional import motor. J Cell Biol 2004; 165:167-173.
72. Matouschek A, Pfanner N, Voos W. Protein unfolding by mitochondria. The Hsp70 import motor. EMBO Rep 2000; 1:404-410.
73. Neupert W, Brunner M. The protein import motor of mitochondria. Nat Rev Mol Cell Biol 2002; 3:555-565.
74. Pfanner N, Geissler A. Versatility of the mitochondrial protein import machinery. Nat Rev Mol Cell Biol 2001; 2:339-349.
75. Mokranjac D, Sichting M, Neupert W et al. Tim14, a novel key component of the import motor of the TIM23 protein translocase of mitochondria. EMBO J 2003; 22:4945-4956.
76. Truscott KN, Voos W, Frazier AE et al. A J-protein is an essential subunit of the presequence translocase-associated protein import motor of mitochondria. J Cell Biol 2003; 163:707-713.
77. Kozany C, Mokranjac D, Sichting M et al. The J domain-related cochaperone Tim16 is a constituent of the mitochondrial TIM23 preprotein translocase. Nat Struct Mol Biol 2004; 11:234-241.
78. Frazier AE, Dudek J, Guiard B et al. Pam16 has an essential role in the mitochondrial protein import motor. Nat Struct Mol Biol 2004; 11:226-233.
79. Gakh O, Cavadini P, Isaya G. Mitochondrial processing peptidases. Biochim Biophys Acta 2002; 1592:63-77.
80. Koehler CM, Merchant S, Schatz G. How membrane proteins travel across the mitochondrial intermembrane space. Trends Biochem Sci 1999; 24:428-432.
81. Rehling P, Pfanner N, Meisinger C. Insertion of hydrophobic membrane proteins into the inner mitochondrial membrane-a guided tour. J Mol Biol 2003; 326:639-657.
82. Komiya T, Rospert S, Schatz G et al. Binding of mitochondrial precursor proteins to the cytoplasmic domains of the import receptors Tom70 and Tom20 is determined by cytoplasmic chaperones. EMBO J 1997; 16:4267-4275.
83. Endres M, Neupert W, Brunner M. Transport of the ADP/ATP carrier of mitochondria from the TOM complex to the TIM22.54 complex. EMBO J 1999; 18:3214-3221.

84. Curran SP, Leuenberger D, Schmidt E et al. The role of the Tim8p-Tim13p complex in a conserved import pathway for mitochondrial polytopic inner membrane proteins. J Cell Biol 2002; 158:1017-1027.
85. Stan T, Brix J, Schneider-Mergener J et al. Mitochondrial protein import: Recognition of internal import signals of BCS1 by the TOM complex. Mol Cell Biol 2003; 23:2239-2250.
86. Luciano P, Vial S, Vergnolle MA et al. Functional reconstitution of the import of the yeast ADP/ATP carrier mediated by the TIM10 complex. EMBO J 2001; 20:4099-4106.
87. Curran SP, Leuenberger D, Oppliger W et al. The Tim9p-Tim10p complex binds to the transmembrane domains of the ADP/ATP carrier. EMBO J 2002; 21:942-953.
88. Vasiljev A, Ahting U, Nargang FE et al. Reconstituted TOM core complex and Tim9/Tim10 complex of mitochondria are sufficient for translocation of the ADP/ATP carrier across membranes. Mol Biol Cell 2004; 15:1445-1458.
89. Koehler CM, Leuenberger D, Merchant S et al. Human deafness dystonia syndrome is a mitochondrial disease. Proc Natl Acad Sci USA 1999; 96:2141-2146.
90. Kurz M, Martin H, Rassow J et al. Biogenesis of Tim proteins of the mitochondrial carrier import pathway: Differential targeting mechanisms and crossing over with the main import pathway. Mol Biol Cell 1999; 10:2461-2474.
91. Jensen RE, Dunn CD. Protein import into and across the mitochondrial inner membrane: Role of the TIM23 and TIM22 translocons. Biochim Biophys Acta 2002; 1592:25-34.
92. Davis AJ, Sepuri NB, Holder J et al. Two intermembrane space TIM complexes interact with different domains of Tim23p during its import into mitochondria. J Cell Biol 2000; 150:1271-1282.
93. Paschen SA, Rothbauer U, Kaldi K et al. The role of the TIM8-13 complex in the import of Tim23 into mitochondria. EMBO J 2000; 19:6392-6400.
94. Koehler CM, Jarosch E, Tokatlidis K et al. Import of mitochondrial carriers mediated by essential proteins of the intermembrane space. Science 1998; 279:369-373.
95. Koehler CM, Merchant S, Oppliger W et al. Tim9p, an essential partner subunit of Tim10p for the import of mitochondrial carrier proteins. EMBO J 1998; 17:6477-6486.
96. Sirrenberg C, Endres M, Fölsch H et al. Zinc finger-like proteins Tim10/Mrs11p and Tim12/Mrs5p mediating import of carrier proteins into mitochondria. Nature 1998; 391:912-915.
97. Adam A, Endres M, Sirrenberg C et al. Tim9, a new component of the TIM22.54 translocase in mitochondria. EMBO J 1999; 18:313-319.
98. Tokatlidis K, Schatz G. Biogenesis of mitochondrial inner membrane proteins. J Biol Chem 1999; 274:35285-35288.
99. Rehling P, Model K, Brandner K et al. Protein insertion into the mitochondrial inner membrane by a twin-pore translocase. Science 2003; 299:1747-1751.
100. Kuhn A, Stuart R, Henry R et al. The Alb3/Oxa1/YidC protein family: Membrane-localized chaperones facilitating membrane protein insertion? Trends Cell Biol 2003; 13:510-516.
101. Hell K, Herrmann JM, Pratje E et al. Oxa1p, an essential component of the novel N-tail protein export machinery in mitochondria. Proc Natl Acad Sci USA 1998; 95:2250-2255.
102. Hell K, Neupert W, Stuart RA. Oxa1p acts as a general membrane insertion machinery for proteins encoded by mitochondrial DNA. EMBO J 2001; 20:1281-1288.
103. Szyrach G, Ott M, Bonnefoy N et al. Ribosome binding to the Oxa1 complex facilitates cotranslational protein insertion in mitochondria. EMBO J 2003; 22:6448-6457.
104. Jia L, Dienhart M, Schramp M et al. Yeast Oxa1 interacts with mitochondrial ribosomes: The importance of the C-terminal region of Oxa1. EMBO J 2003; 22:6438-6447.

CHAPTER 10

Import of Proteins into Peroxisomes

Sven Thoms* and Ralf Erdmann

Abstract

Peroxisomes are organelles equipped with enzymes for lipid metabolism and hydrogen-peroxide-based respiration. Though many details of their metabolism are understood today, basic aspects concerning their biogenesis, including translocation of peroxisomal proteins into and through the peroxisomal membrane, still remain unknown. Nevertheless, the past years have brought forth a wealth of detailed information on the proteins required for proper biogenesis of peroxisomes. This review focuses on the basic principles and on recent developments in the field of peroxisome biogenesis. More comprehensive or specialized reviews can be found in the reference list.[1-6]

Introduction

Peroxisomes are seemingly simple cellular organelles present in virtually all eukaryotes. They are surrounded by a single lipid bilayer and appear as spherical organelles, or, in a more detailed morphological analysis, as a reticular network.[7]

Peroxisomes are endowed with enzymes for fatty acid β-oxidation, hydrogen peroxide-producing oxidases and catalase, serving roles in respiration and defence against oxygen stress. In yeast, peroxisomes are involved in the biosynthesis of lysine and are the only site for β-oxidation. Peroxisomes in mammals are additionally involved in plasmalogen biosynthesis. In plants, peroxisomes are the sites of photorespiration and the glyoxylate cycle. In trypanosomes, glycolysis is exclusively localized to their peroxisomes.

The importance of the study of peroxisomes is underscored by the existence of inborn human defects of peroxisome function. Many of these disorders have devastating effects on the life of the patient. Disorders of peroxisome function can be classified either as single enzyme disorders (such as phytanoyl-CoA hydroxylase deficiency and X-linked adrenoleucodystrophy) or as peroxisomal biogenesis disorders (PBDs). The latter group comprises diseases of the Zellweger spectrum, including Zellweger syndrome, neonatal adrenoleucodystrophy and infantile Refsum disease. PBDs are caused by defective transport of peroxisomal matrix or membrane proteins and can, therefore, be regarded as protein targeting diseases. For recent reviews, see references 8-10.

Proteins required for the biogenesis of peroxisomes, collectively called peroxins, are encoded by PEX genes. The first of these proteins have been identified in yeast and CHO cells

*Corresponding Author—Sven Thoms, Institute for Physiological Chemistry, University of Bochum Universitätsstrasse 150, 44780, Bochum, Germany.
Email: sven.thoms@ruhr-uni-bochum.de

Protein Movement Across Membranes, edited by Jerry Eichler. ©2005 Eurekah.com and Springer Science+Business Media.

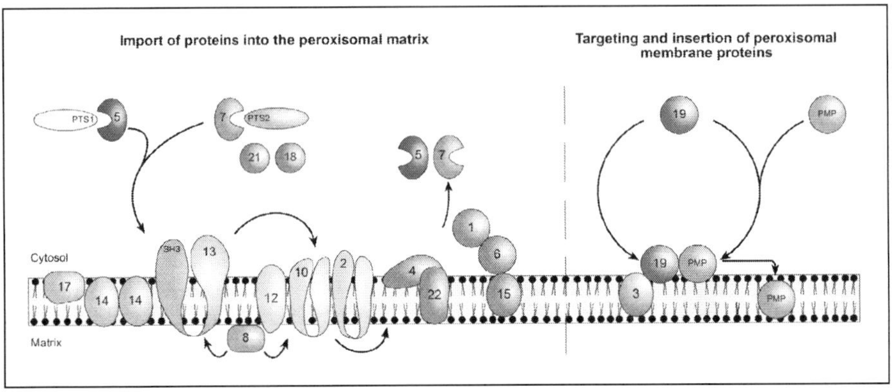

Figure 1. Models of the topogenesis of peroxisomal matrix and membrane proteins.

through genetic screens.[11-13] Today, proteomic studies[14,15] and in vitro systems[16-19] add to the tools available for understanding mechanisms of peroxisome targeting. The high degree of evolutionary conservation in peroxins across species has enabled a unified nomenclature for peroxins.[20] As of 2004, the count goes up to 32 peroxins, with about 24 being highly conserved amongst species.

Conceptually, import of soluble proteins into peroxisomes has been subdivided into the following steps: First, binding of the peroxisomal targeting signals (PTS) of import substrates to their respective cytosolic receptors Pex5p or Pex7p; second, binding of these receptor complexes to a docking complex at the peroxisomal membrane; third, dissociation of the receptor-cargo complexes and translocation of the cargo proteins across the peroxisomal membrane; fourth, recycling or removal of the receptors. We assume this conceptualization reflects a protein cascade of temporal and/or spatial steps in matrix protein import[21] (Fig. 1).

The targeting of membrane proteins to the peroxisomal membrane is less well understood. It is known, however, that the farnesylated protein Pex19p and the peroxisomal membrane protein Pex3p as well as a lesser-conserved Pex16p protein play essential roles in this process. We will address the question of matrix protein import and PMP (peroxisomal membrane protein) insertion in turn.

Matrix Protein Import

Proteins that are to be targeted to the matrix of peroxisomes are encoded in the nucleus, synthesised on free ribosomes and subsequently imported into the peroxisome. Matrix proteins contain a C- or N-terminal peroxisomal targeting signal.

After the discovery of the peroxisomal targeting of firefly luciferase,[22] the C-terminal tripeptide serine-lysine-leucine (SKL) was shown to be sufficient for directing a protein to peroxisomes[23] and still represents an efficient peroxisome targeting signal (PTS) for matrix protein import into peroxisomes in all species studied so far. PTS1 is described as a C-terminal tripeptide of the consensus [SAC] [KRH] [LM]. Masking of the C-terminal PTS, e.g., by the addition of GFP, abolishes peroxisomal localization,[24] indicating that the signal has to be located at the very C-terminus.

Interestingly, the first functional viral PTS1 was recently identified in the VP4 protein of rotavirus,[25] a non-enveloped virus that causes gastroenteritis, killing 440,000 children annually.[26] It will be fascinating to learn how viral pathogenicity is related to peroxisome function. Putative PTSs have been identified in a number of viral proteins.[27]

Pex5p is the soluble receptor for proteins bearing a PTS1. Pex5p consists of two domains. The N-terminal half contains WxxxF-repeats, yet otherwise has few conserved residues. This half is involved in targeting by binding to Pex14p and in a number of other protein-protein interactions.[28,29] The C-terminal domain contains the tetratricopeptide repeats (TPR), which are involved in binding the PTS1 (see ref. 30). TPRs are α-helical 34 amino acid repeats involved in protein-protein interactions. The crystal structure of the C-terminus of human Pex5p in a complex with a PTS1-peptide reveals that the two clusters of three TPRs almost completely engulf the targeting peptide.[31] The seventh TPR is part of the hinge region between the two clusters.

A second matrix protein peroxisomal targeting signal (PTS2) was first discovered in thiolase.[32] PTS2 is found in considerably fewer proteins in all organisms. In *Saccharomyces cerevisiae*, for example, thiolase is the only known PTS2 protein, and in *Caenorhabditis elegans*, the PTS2 pathway seems to be completely absent.[33] PTS2 is usually located within the first 20 amino acids of the protein and has recently been redefined as [RK][LVIQ]XX[LVIHQ][LSGAK]X[HQ][LAF], covering virtually all PTS2 variants.[34] Pex8p in yeast is a rare example of a protein carrying both PTS1 and PTS2.[35]

PTS2-bearing proteins are recognised by their cytosolic receptor Pex7p.[36] In lower eukaryotes, PTS2 targeting requires accessory proteins of the Pex20p-family (Pex18p and Pex21p in *S. cerevisiae*,[37] Pex20p in *Yarrowia lipolytica*[38] or *Neurospora crassa*[39]). In mammals, Pex7p binds to the longer form of the two splicing variants of Pex5p,[40,41] making PTS2 import eventually dependent on Pex5p. In agreement with their common function, there is sequence similarity between the extension in Pex5L(ong)p and Pex20p.[42,43]

The 'shuttle model' describes Pex5p and Pex7p as cycling receptors that bind their import cargo in the cytosol and transport it to the docking complex at the peroxisomal membrane.[36,44] The 'extended shuttle model' accounts for entry of Pex5p into peroxisomes.[45,46] Support for the shuttle model has also come from recent data showing a specific requirement of the N-terminus of Pex5p for its recycling.[19]

The docking complex in yeast includes peroxins 13, 14 and 17. These peroxins are thought to provide the contact platform for the receptor complexes at the peroxisomal membrane. Pex14p is a peroxisomal membrane protein[47] which binds to Pex5p and, in yeast, is dependent on Pex13p for its localization at the peroxisomal membrane.[48]

Pex13p is a peroxisomal membrane protein with both its termini exposed to the cytosol. The C-terminus contains a Src-homology 3 (SH3) domain, which binds Pex14p and Pex5p.[49-51] The structure of the SH3 domain has recently been solved by X-ray crystallography[52] and by NMR spectroscopy, revealing separate binding sites for the type II SH3 ligand Pex14p and the non-PXXP protein, Pex5p.[53] The docking complex, by its various binding sites, is thought to provide a template for sequential interaction of receptor-cargo complexes with the docking site and subsequent recycling of the receptors.[54]

As has recently been demonstrated, the PTS1 import receptor Pex5p is modified by ubiquitination.[55] In cells deficient in the ubiquitin conjugating enzyme Pex4p/Ubc10p[56] or its membrane receptor Pex22p, Pex5p is modified by mono-ubiquitin, whereas in cells deficient in the AAA peroxins (see below), poly-ubiquitin chains predominate. Ubiquitination of Pex5p itself is dependent on the ubiquitin conjugation enzyme Ubc4p and takes place at the peroxisomal membrane after the docking phase.[55] It seems likely that ubiquitination of Pex5p and Pex18p[57] are essential steps in peroxisome biogenesis.

Pex2p, Pex10p and Pex12p are PMPs with C_3HC_4 (RING) finger domains. The three peroxins interact with each other, and Pex10p and Pex12p can also bind Pex5p. This latter interaction is thought to be required further downstream of docking.[58] RING domain proteins are often mediators of ubiquitin ligases, so it is an appealing speculation that these peroxins may be involved in ubiquitination of Pex5p and Pex18p.

The interaction of the docking complex with the RING complex requires intraperoxisomal Pex8p, which appears to be an essential coordinator of the docking and the RING complex.[59] Pex8p alone can also interact with the PTS1 import receptor.[60]

The peroxins Pex1p and Pex6p are ATPases of the AAA-type (ATPases associated with various cellular activities). AAA proteins, originally defined after the identification of Pex1p,[11] constitute a growing family of hexameric proteins including members such as NSF or p97/ Cdc48p, involved in processing protein complexes and in membrane fusion. (see refs. 61, 62 for review.) Pex1p and Pex6p interact with each other. There is a dispute as to whether Pex1p and Pex6p are associated with peroxisomes or different cellular structures,[63] or peroxisomal precursors.[64] Recently, Pex15p has been identified as a peroxisomal membrane receptor for Pex6p in yeast.[65] The membrane receptor for Pex6p in mammalian cells has been termed Pex26p.[66]

Pex1p and Pex6p are the only ATPases among the peroxins. It is, therefore, likely that these proteins confer the ATP requirement for protein import into peroxisomes. Pex1p and Pex6p could be responsible for (i) separating the import receptor from its cargo immediately before import, or (ii) for recycling of the import receptor. It has also been suggested that the AAA peroxins (iii) are responsible for peroxisome fusion[67] or serve as (iv) defining components of lipid ferries to the peroxisome.[3]

One of the most striking features of the import machinery for matrix proteins is its ability to import folded proteins and protein complexes.[68] Even gold particles, when coated with a PTS, can access the peroxisome.[69] Furthermore, proteins without a PTS can be piggy-back transported into the peroxisome together with PTS-containing subunits after their oligomerization in the cytoplasm.[70] An especially remarkable import substrate might be the Nef protein from the HIV-1. This non-PTS protein is transported together with human thioesterase II into peroxisomes, so that Nef is localized to peroxisomes.[71] A stably folded protein conformation, however, is not a precondition for import into peroxisomes.[72]

Hsp70-homologues escort proteins while or after they are released from the ribosome[73] and keep them in an extended conformation which facilitates transfer into cellular organelles. In spite of the ability of peroxisomes to import folded and oligomerized proteins, there is evidence for the involvement of Hsp70 proteins in matrix protein import.[74-76] Hsp70 proteins might regulate the binding of the import receptor to its cargo.[77] Cochaperones of Hsp70 that belong to the J domain protein family regulate Hsp70 proteins. J domain proteins have been identified at the peroxisomal membrane[78,79] and a yeast DnaJ protein is needed for matrix import in peroxisomes in an early, yet post-translational phase.[80]

Cargo association, together with receptor oligomerization,[81] has stimulated the idea of high molecular weight cargo-receptor complexes forming before import. These complexes have been termed 'preimplexes", short for 'preimport complexes'.[6] Preimplexes could 'configurate' import substrates before their import at the peroxisomal membrane. Their existence, however, is still unproven.

There is now detailed knowledge on the protein factors involved in matrix protein import. The fundamental mechanism, however, of how proteins cross the peroxisomal membrane has yet to be discovered. Three models have been suggested.[82] One suggests a permanent (static) pore in the peroxisomal membrane which allows the entry and exit of cargo. The dynamic pore model, in contrast, suggests that the pore components are recruited only after docking. This model envisions a signal-assembled translocon[83] which would consist of the docking and RING complexes, or parts thereof. The third model suggests protein import by membrane internalization akin to endocytosis.[82] Membranes would have to invaginate into the peroxisome, introducing intracellular vesicles into the peroxisomes.

PMP Import and the Origin of Peroxisomes

The membrane of peroxisomes contains proteins required for metabolite transport and peroxisome biogenesis. Like matrix proteins, most PMPs are believed to be post-translationally imported into peroxisomes.[84] PMP topogenesis is independent of matrix protein import,[49] although matrix proteins require the above-mentioned PMPs for their import. Cells from Zellweger patients were found to contain ghosts, i.e., empty peroxisomal membranes that lacked most, if not all, matrix proteins.[85]

The proteins Pex3p, Pex19p and Pex16p are required for the biogenesis of the peroxisomal membrane. The absence or mutation of one of these proteins causes the absence of peroxisomes and formation of peroxisomal ghosts in yeast[86] as well as in man.[87] Whether peroxisomal ghosts solely represent aberrant vesicles of defective peroxisome biogenesis, or if they are related to a somewhat mysterious 'proto-peroxisome' is currently under debate.

Pex19p[88] is a predominantly cytosolic protein that interacts with several PMPs and has been suggested to act as a soluble import receptor for PMPs[89,90] and a chaperone that either supports integration of PMPs into the membrane[91] or the assembly of PMP complexes in the membrane[92] (Fig. 1). Recent evidence, however, is contradictory in that (1) Pex19p has been demonstrated to capture PMPs in the cytosol and to target them to the peroxisomal membrane[93] and (2) peroxisomal targeting of some PMPs is independent of Pex19p binding.[94] Pex19p is a farnesylated protein with a highly conserved farnesylation motif (C-terminal CaaX), which is essential for proper Pex19p function.[88] The farnesyl group could be involved in membrane and/or substrate binding. Pex3p has recently been described as a receptor or docking factor for Pex19p.[93]

In comparison to matrix protein targeting signals, membrane targeting signals (mPTS) are more difficult to identify. Pex3p, for example, has its transmembrane domain (TMD) at the N-terminus, close to or overlapping with the mPTS.[95] Similarly, the targeting information of APX is thought to be constituted of the basic composition of a C-terminal stretch, rather than by a specific sequence.[96] PMPs with a single TMD seem to carry a set of positively-charged amino acids next to a hydrophobic patch or the TMD.[97,98] In proteins with several TMDs, cooperation of several regions seems to be required for membrane localization.[99] A recent attempt to identify a more universal mPTS has been made by using peptide scans based on Pex19p binding sites in Pex11p and Pex13p. This approach has led to the identification of a 'prediction matrix' for Pex19p binding and membrane insertion.[100] The new data demonstrate that the Pex19p binding site, in conjugation with an adjacent transmembrane domain, functions as an mPTS for most PMPs. Moreover, these data underline the functions of Pex19p as a signal sequence receptor for peroxisomal membrane proteins.[100]

Pex19p, however, has also been associated with cellular effects that might not directly relate to peroxisomes. These include an interaction with the renal type IIa sodium-dependent phosphate cotransporter[101] and down-regulation of p53 by interaction with the tumor suppressor protein ARF in mouse,[102] but not in humans.[103]

The idea that peroxisomes bud off from the ER dates back to the early days of peroxisome research[104,105] and has recently been supported by work showing that peroxisomes originate in the vicinity of the ER.[106] Originally, the discovery that most peroxisomal proteins are synthesised on free ribosomes has led to a rejection of ER-based scenarios in favour of a model suggesting that peroxisomes proliferate by 'growth and division'.[84] In line with this model, it has been shown that inactivation of parts of the sec-dependent secretion pathway does not affect peroxisome biogenesis,[107,108] arguing in favor of ER-independent models. ER-independent models would rely on 'proto-peroxisomes' as a basis for peroxisome biogenesis. These are either 'back-up' peroxisomes, derived from peroxisomes to ensure their proliferation, or they are derived from other cellular membranes. On the other hand, there is now good evidence that certain peroxisomal proteins can reach the peroxisome via the ER.[109,110]

In conclusion, we have to envisage a model of peroxisome biogenesis in which a given peroxisome has received its proteins from three different sources: (1) by import of cytosolic proteins, (2) by transport of proteins through the ER, and (3) from another peroxisome by division.

Peroxins are well-conserved throughout evolution, however, we know very little on how the peroxisomes have themselves evolved,[111] and to what extent peroxins and the mechanisms of peroxisome biogenesis have been conserved throughout evolution. Obviously, more detailed knowledge on the evolution of peroxisomes would help to understand their biogenesis and vice versa. The peroxisome was one of the last cellular organelles to be discovered. It may also be the last whose mechanism of biogenesis is to be deciphered.

Acknowledgements

We thank Mathula Thangarajh, Hanspeter Rottensteiner and Wolfgang Schliebs for reading of the manuscript. This work is supported by the Deutsche Forschungsgemeinschaft (ER178/2-4) and the Fonds der Chemischen Industrie.

References

1. Eckert JH, Erdmann R. Peroxisome biogenesis. Rev Physiol Biochem Pharmacol 2003; 147:75-121.
2. Brown LA, Baker A. Peroxisome biogenesis and the role of protein import. J Cell Mol Med 2003; 7:388-400.
3. Lazarow PB. Peroxisome biogenesis: Advances and conundrums. Curr Opin Cell Biol 2003; 15:489-497.
4. Titorenko VI, Rachubinski RA. The peroxisome: Orchestrating important developmental decisions from inside the cell. J Cell Biol 2004; 164:641-645.
5. Veenhuis M, Kiel JA, Van Der Klei IJ. Peroxisome assembly in yeast. Microsc Res Tech 2003; 61:139-150.
6. Gould SJ, Collins CS. Opinion: Peroxisomal-protein import: Is it really that complex? Nat Rev Mol Cell Biol 2002; 3:382-389.
7. Schrader M, King SJ, Stroh TA et al. Real time imaging reveals a peroxisomal reticulum in living cells. J Cell Sci 2000; 113:3663-3671.
8. Baes M, Van Veldhoven PP. Lessons from knockout mice. I: Phenotypes of mice with peroxisome biogenesis disorders. Adv Exp Med Biol 2003; 544:113-122.
9. Weller S, Gould SJ, Valle D. Peroxisome biogenesis disorders. Annu Rev Genomics Hum Genet 2003; 4:165-211.
10. Wanders RJ. Metabolic and molecular basis of peroxisomal disorders: A review. Am J Med Genet 2004; 126A:355-375.
11. Erdmann R, Wiebel FF, Flessau A et al. PAS1, a yeast gene required for peroxisome biogenesis, encodes a member of a novel family of putative ATPases. Cell 1991; 64:499-510.
12. Tsukamoto T, Yokota S, Fujiki Y. Isolation and characterization of Chinese hamster ovary cell mutants defective in assembly of peroxisomes. J Cell Biol 1990; 110:651-660.
13. Tsukamoto T, Miura S, Fujiki Y. Restoration by a 35K membrane protein of peroxisome assembly in a peroxisome-deficient mammalian cell mutant. Nature 1991; 350:77-81.
14. Kikuchi M, Hatano N, Yokota S et al. Proteomic analysis of rat liver peroxisome: Presence of peroxisome-specific isozyme of lon protease. J Biol Chem 2004; 279:421-428.
15. Schäfer H, Nau K, Sickmann A et al. Identification of peroxisomal membrane proteins of Saccharomyces cerevisiae by mass spectrometry. Electrophoresis 2001; 22:2955-2968.
16. Pool MR, Lopez-Huertas E, Baker A. Characterization of intermediates in the process of plant peroxisomal protein import. EMBO J 1998; 17:6854-6862.
17. Baker A, Charlton W, Johnson B et al. Biochemical and molecular approaches to understanding protein import into peroxisomes. Biochem Soc Trans 2000; 28:499-504.
18. Gouveia AM, Guimaraes CP, Oliveira ME et al. Characterization of the peroxisomal cycling receptor Pex5p import pathway. Adv Exp Med Biol 2003; 544:213-220.

19. Costa-Rodrigues J, Carvalho AF, Gouveia AM et al. The N-terminus of the peroxisomal cycling receptor, Pex5p, is required for redirecting the peroxisome-associated peroxin back to the cytosol. J Biol Chem 2004; 279:15034-15041.
20. Distel B, Erdmann R, Gould SJ et al. A unified nomenclature for peroxisome biogenesis factors. J Cell Biol 1996; 135:1-3.
21. Holroyd C, Erdmann R. Protein translocation machineries of peroxisomes. FEBS Lett 2001; 501:6-10.
22. Keller GA, Gould S, DeLuca M et al. Firefly luciferase is targeted to peroxisomes in mammalian cells. Proc Natl Acad Sci USA 1987; 84:3264-3268.
23. Gould SJ, Keller GA, Hosken N et al. A conserved tripeptide sorts proteins to peroxisomes. J Cell Biol 1989; 108:1657-1664.
24. Huh WK, Falvo JV, Gerke LC et al. Global analysis of protein localization in budding yeast. Nature 2003; 425:686-691.
25. Mohan KV, Som I, Atreya CD. Identification of a Type 1 peroxisomal targeting signal in a viral protein and demonstration of its targeting to thc organcllc. J Virol 2002; 76:2543-2547.
26. Parashar UD, Hummelman EG, Bresee JS et al. Global illness and deaths caused by rotavirus disease in children. Emerg Infect Dis 2003; 9:565-572.
27. Mohan KV, Atreya CD. Novel organelle-targeting signals in viral proteins. Bioinformatics 2003; 19:10-13.
28. Otera H, Setoguchi K, Hamasaki M et al. Peroxisomal targeting signal receptor Pex5p interacts with cargoes and import machinery components in a spatiotemporally differentiated manner: Conserved Pex5p WXXXF/Y motifs are critical for matrix protein import. Mol Cell Biol 2002; 22:1639-1655.
29. Saidowsky J, Dodt G, Kirchberg K et al. The Di-aromatic pentapeptide repeats of the human peroxisome import receptor PEX5 are separate high-affinity binding sites for the peroxisomal membrane protein PEX14. J Biol Chem 2001; 276:34524-34529.
30. Klein AT, Barnett P, Bottger G et al. Recognition of the peroxisomal targeting signal type 1 by the protein import receptor Pex5p. J Biol Chem 2001; 11.
31. Gatto GJJ, Geisbrecht BV, Gould SJ et al. Peroxisomal targeting signal-1 recognition by the TPR domains of human PEX5. Nat Struct Biol 2000; 7:1091-1095.
32. Swinkels BW, Gould SJ, Bodnar AG et al. A novel, cleavable peroxisomal targeting signal at the amino-terminus of the rat 3-ketoacyl-CoA thiolase. EMBO J 1991; 10:3255-3262.
33. Motley AM, Hettema EH, Ketting R et al. Caenorhabditis elegans has a single pathway to target matrix proteins to peroxisomes. EMBO Rep 2000; 1:40-46.
34. Petriv OI, Tang L, Titorenko VI et al. A new definition for the consensus sequence of the peroxisome targeting signal type 2. J Mol Biol 2004; 341:119-134.
35. Waterham HR, Titorenko VI, Haima P et al. The Hansenula polymorpha PER1 gene is essential for peroxisome biogenesis and encodes a peroxisomal matrix protein with both carboxy- and amino-terminal targeting signals. J Cell Biol 1994; 127:737-749.
36. Marzioch M, Erdmann R, Veenhuis M et al. PAS7 encodes a novel yeast member of the WD-40 protein family essential for import of 3-oxoacyl-CoA thiolase, a PTS2-containing protein, into peroxisomes. EMBO J 1994; 13:4908-4918.
37. Stein K, Schell-Steven A, Erdmann R et al. Interactions of Pex7p and Pex18p/Pex21p with the peroxisomal docking machinery: Implications for the first steps in PTS2 protein import. Mol Cell Biol 2002; 22:6059-6069.
38. Titorenko VI, Smith JJ, Szilard RK et al. Pex20p of the yeast Yarrowia lipolytica is required for the oligomerization of thiolase in the cytosol and for its targeting to the peroxisome. J Cell Biol 1998; 142:403-420.
39. Sichting M, Schell-Steven A, Prokisch H et al. Pex7p and Pex20p of Neurospora crassa function together in PTS2-dependent protein import into peroxisomes. Mol Biol Cell 2003; 14:810-821.
40. Otera H, Harano T, Honsho M et al. The mammalian peroxin Pex5pL, the longer isoform of the mobile peroxisome targeting signal (PTS) type 1 transporter, translocates the Pex7p-PTS2 protein complex into peroxisomes via its initial docking site, Pex14p. J Biol Chem 2000; 275:21703-21714.

41. Matsumura T, Otera H, Fujiki Y. Disruption of the interaction of the longer isoform of Pex5p, Pex5pL, with Pex7p abolishes peroxisome targeting signal type 2 protein import in mammals. Study with a novel Pex5-impaired Chinese hamster ovary cell mutant. J Biol Chem 2000; 275:21715-21721.
42. Einwächter H, Sowinski S, Kunau WH et al. Yarrowia lipolytica Pex20p, Saccharomyces cerevisiae Pex18p/Pex21p and mammalian Pex5pL fulfil a common function in the early steps of the peroxisomal PTS2 import pathway. EMBO Rep 2001; 2:1035-1039.
43. Dodt G, Warren D, Becker E et al. Domain mapping of human PEX5 reveals functional and structural similarities to Saccharomyces cerevisiae Pex18p and Pex21p. J Biol Chem 2001; 276:41769-41781.
44. Dodt G, Gould SJ. Multiple PEX genes are required for proper subcellular distribution and stability of Pex5p, the PTS1 receptor: Evidence that PTS1 protein import is mediated by a cycling receptor. J Cell Biol 1996; 135:1763-1774.
45. Dammai V, Subramani S. The human peroxisomal targeting signal receptor, Pex5p, is translocated into the peroxisomal matrix and recycled to the cytosol. Cell 2001; 105:187-196.
46. Kunau W. Peroxisomes: The extended shuttle to the peroxisome matrix. Curr Biol 2001; 11:R659-R662.
47. Albertini M, Rehling P, Erdmann R et al. Pex14p, a peroxisomal membrane protein binding both receptors of the two PTS-dependent import pathways. Cell 1997; 89:83-92.
48. Girzalsky W, Rehling P, Stein K et al. Involvement of Pex13p in Pex14p localization and peroxisomal targeting signal 2-dependent protein import into peroxisomes. J Cell Biol 1999; 144:1151-1162.
49. Erdmann R, Blobel G. Identification of Pex13p a peroxisomal membrane receptor for the PTS1 recognition factor. J Cell Biol 1996; 135:111-121.
50. Elgersma Y, Kwast L, Klein A et al. The SH3 domain of the Saccharomyces cerevisiae peroxisomal membrane protein Pex13p functions as a docking site for Pex5p, a mobile receptor for the import of PTS1 containing proteins. J Cell Biol 1996; 135:97-109.
51. Gould SJ, Kalish JE, Morrell JC et al. Pex13p is an SH3 protein of the peroxisome membrane and a docking factor for the predominantly cytoplasmic PTS1 receptor. J Cell Biol 1996; 135:85-95.
52. Douangamath A, Filipp FV, Klein AT et al. Topography for independent binding of alpha-helical and PPII-helical ligands to a peroxisomal SH3 domain. Mol Cell 2002; 10:1007-1017.
53. Pires JR, Hong X, Brockmann C et al. The ScPex13p SH3 domain exposes two distinct binding sites for Pex5p and Pex14p. J Mol Biol 2003; 326:1427-1435.
54. Reguenga C, Oliveira ME, Gouveia AM et al. Characterization of the mammalian peroxisomal import machinery: Pex2p, Pex5p, Pex12p, and Pex14p are subunits of the same protein assembly. J Biol Chem 2001; 276:29935-29942.
55. Platta HW, Girzalsky W, Erdmann R. Ubiquitination of the peroxisomal import receptor Pex5p. Biochem J 2004; 384:37-45.
56. Wiebel FF, Kunau W-H. The PAS2 protein essential for peroxisome biogenesis is related to ubiquitin-conjugating enzymes. Nature 1992; 359:73-76.
57. Purdue PE, Lazarow PB. Pex18p is constitutively degraded during peroxisome biogenesis. J Biol Chem 2001; 276:47684-47689.
58. Collins CS, Kalish JE, Morrell JC et al. The peroxisome biogenesis factors Pex4p, Pex22p, Pex1p, and Pex6p act in the terminal steps of peroxisomal matrix protein import. Mol Cell Biol 2000; 20:7516-7526.
59. Agne B, Meindl NM, Niederhoff K et al. Pex8p. An intraperoxisomal organizer of the peroxisomal import machinery. Mol Cell 2003; 11:635-646.
60. Rehling P, Skaletz-Rorowski A, Girzalsky W et al. Pex8p, an intraperoxisomal peroxin of Saccharomyces cerevisiae required for protein transport into peroxisomes binds the PTS1 receptor pex5p. J Biol Chem 2000; 275:3593-3602.
61. Ogura T, Wilkinson AJ. AAA+ superfamily ATPases: Common structure-diverse function. Genes Cells 2001; 6:575-597.
62. Lupas AN, Martin J. AAA proteins. Curr Opin Struct Biol 2002; 12:746-753.
63. Faber KN, Heyman JA, Subramani S. Two AAA family peroxins, PpPex1p and PpPex6p, interact with each other in an ATP-dependent manner and are associated with different subcellular membranous structures distinct from peroxisomes. Mol Cell Biol 1998; 18:936-943.

64. Titorenko VI, Rachubinski RA. The life cycle of the peroxisome. Nat Rev Mol Cell Biol 2001; 2:357-368.
65. Birschmann I, Stroobants AK, Van Den Berg M et al. Pex15p of Saccharomyces cerevisiae provides a molecular basis for recruitment of the AAA peroxin Pex6p to peroxisomal membranes. Mol Biol Cell 2003; 14:2226-2236.
66. Matsumoto N, Tamura S, Fujiki Y. The pathogenic peroxin Pex26p recruits the Pex1p-Pex6p AAA ATPase complexes to peroxisomes. Nat Cell Biol 2003; 5:454-460.
67. Titorenko VI, Rachubinski RA. Peroxisomal membrane fusion requires two AAA family ATPases, Pex1p and Pex6p. J Cell Biol 2000; 150:881-886.
68. Bellion E, Goodman JM. Proton ionophores prevent assembly of a peroxisomal protein. Cell 1987; 48:165-173.
69. Walton PA, Hill PE, Subramani S. Import of stably folded proteins into peroxisomes. Mol Biol Cell 1995; 6:675-683.
70. Yang X, Purdue PE, Lazarow PB. Eci1p uses a PTS1 to enter peroxisomes: Either its own or that of a partner, Dci1p. Eur J Cell Biol 2001; 80:126-138.
71. Cohen GB, Rangan VS, Chen BK et al. The human thioesterase II protein binds to a site on HIV-1 Nef critical for CD4 down-regulation. J Biol Chem 2000; 275:23097-23105.
72. Brocard CB, Jedeszko C, Song HC et al. Protein structure and import into the peroxisomal matrix. Traffic 2003; 4:74-82.
73. Craig EA, Eisenman HC, Hundley HA. Ribosome-tethered molecular chaperones: The first line of defense against protein misfolding? Curr Opin Microbiol 2003; 6:157-162.
74. Walton PA, Wendland M, Subramani S et al. Involvement of 70-kD heat-shock proteins in peroxisomal import. J Cell Biol 1994; 125:1037-1046.
75. Legakis JE, Terlecky SR. PTS2 Protein import into mammalian peroxisomes. Traffic 2001; 2:252-260.
76. Wimmer B, Lottspeich F, van der Klei I et al. The glyoxysomal and plastid molecular chaperones (70-kDa heat shock protein) of watermelon cotyledons are encoded by a single gene. Proc Natl Acad Sci USA 1997; 94:13624-13629.
77. Harano T, Nose S, Uezu R et al. Hsp70 regulates the interaction between the peroxisome targeting signal type 1 (PTS1)-receptor Pex5p and PTS1. Biochem J 2001; 357:157-165.
78. Diefenbach J, Kindl H. The membrane-bound DnaJ protein located at the cytosolic site of glyoxysomes specifically binds the cytosolic isoform 1 of Hsp70 but not other Hsp70 species. Eur J Biochem 2000; 267:746-754.
79. Preisig-Muller R, Muster G, Kindl H. Heat shock enhances the amount of prenylated DnaJ protein at membranes of glyoxysomes. Eur J Biochem 1994; 219:57-63.
80. Hettema EH, Ruigrok CCM, Koerkamp MG et al. The cytosolic DnaJ-like protein Djp1p is involved specifically in peroxisomal protein import. J Cell Biol 1998; 142:421-434.
81. Schliebs W, Saidowsky J, Agianian B et al. Recombinant human peroxisomal targeting signal receptor PEX5. Structural basis for interaction of PEX5 with PEX14. J Biol Chem 1999; 274:5666-5673.
82. McNew JA, Goodman JM. An oligomeric protein is imported into peroxisomes in vivo. J Cell Biol 1994; 127:1245-1257.
83. Schnell DJ, Hebert DN. Protein translocons: Multifunctional mediators of protein translocation across membranes. Cell 2003; 112:491-505.
84. Lazarow PB, Fujiki Y. Biogenesis of peroxisomes. Annu Rev Cell Biol 1985; 1:489-530.
85. Santos MJ, Imanaka T, Shio H et al. Peroxisomal membrane ghosts in Zellweger syndrome-aberrant organelle assembly. Science 1988; 239:1536-1538.
86. Hettema EH, Girzalsky W, van Den Berg M et al. Saccharomyces cerevisiae Pex3p and Pex19p are required for proper localization and stability of peroxisomal membrane proteins. EMBO J 2000; 19:223-233.
87. Honsho M, Hiroshige T, Fujiki Y. The membrane biogenesis peroxin Pex16p. Topogenesis and functional roles in peroxisomal membrane assembly. J Biol Chem 2002; 277:44513-44524.
88. Götte K, Girzalsky W, Linkert M et al. Pex19p, a farnesylated protein essential for peroxisome biogenesis. Mol Cell Biol 1998; 18:616-628.

89. Snyder WB, Koller A, Choy AJ et al. The peroxin Pex19p interacts with multiple, integral membrane proteins at the peroxisomal membrane. J Cell Biol 2000; 149:1171-1178.
90. Sacksteder KA, Jones JM, South ST et al. PEX19 binds multiple peroxisomal membrane proteins, is predominantly cytoplasmic, and is required for peroxisome membrane synthesis. J Cell Biol 2000; 148:931-944.
91. Jones JM, Morrell JC, Gould SJ. PEX19 is a predominantly cytosolic chaperone and import receptor for class 1 peroxisomal membrane proteins. J Cell Biol 2004; 164:57-67.
92. Fransen M, Wylin T, Brees C et al. Human Pex19p binds peroxisomal integral membrane proteins at regions distinct from their sorting sequences. Mol Cell Biol 2001; 21:4413-4424.
93. Fang Y, Morrell JC, Jones JM et al. PEX3 functions as a PEX19 docking factor in the import of class I peroxisomal membrane proteins. J Cell Biol 2004; 164:863-875.
94. Fransen M, Vastiau I, Brees C et al. Potential role for Pex19p in assembly of PTS-receptor docking complexes. J Biol Chem 2004; 279:12615-12624.
95. Baerends RJ, Faber KN, Kram AM et al. A stretch of positively charged amino acids at the N terminus of Hansenula polymorpha Pex3p is involved in incorporation of the protein into the peroxisomal membrane. J Biol Chem 2000; 275:9986-9995.
96. Mullen RT, Trelease RN. The sorting signals for peroxisomal membrane-bound ascorbate peroxidase are within its C-terminal tail. J Biol Chem 2000; 275:16337-16344.
97. Dyer JM, McNew JA, Goodman JM. The sorting sequence of the peroxisomal integral membrane protein PMP47 is contained within a short hydrophilic loop. J Cell Biol 1996; 133:269-280.
98. Wang X, Unruh MJ, Goodman JM. Discrete targeting signals direct Pmp47 to oleate-induced peroxisomes in Saccharomyces cerevisiae. J Biol Chem 2001; 276:10897-10905.
99. Jones JM, Morrell JC, Gould SJ. Multiple distinct targeting signals in integral peroxisomal membrane proteins. J Cell Biol 2001; 153:1141-1150.
100. Rottensteiner H, Kramer A, Lorenzen S et al. Peroxisomal membrane proteins contain common Pex19p-binding sites that are an integral part of their targeting signals. Mol Biol Cell 2004; 15:3406-3417.
101. Ito M, Iidawa S, Izuka M et al. Interaction of a farnesylated protein with renal type IIa Na/Pi cotransporter in response to parathyroid hormone and dietary phosphate. Biochem J 2004; 377:607-616.
102. Sugihara T, Kaul SC, Kato J et al. Pex19p dampens the p19ARF-p53-p21WAF1 tumor suppressor pathway. J Biol Chem 2001; 276:18649-18652.
103. Wadhwa R, Sugihara T, Hasan MK et al. A major functional difference between the mouse and human ARF tumor suppressor proteins. J Biol Chem 2002; 277:36665-36670.
104. Novikoff AB, Shin W-Y. The endoplasmic reticulum in the Golgi zone and its relations to microbodies, Golgi apparatus and autophagic vacuoles in rat liver cells. J Microsc 1964; 3:187-206.
105. Beevers H. Microbodies in higher plants. Annu Rev Plant Physiol 1979; 30:159-193.
106. Tabak HF, Murk JL, Braakman I et al. Peroxisomes start their life in the endoplasmic reticulum. Traffic 2003; 4:512-518.
107. South ST, Baumgart E, Gould SJ. Inactivation of the endoplasmic reticulum protein translocation factor, Sec61p, or its homolog, Ssh1p, does not affect peroxisome biogenesis. Proc Natl Acad Sci USA 2001; 98:12027-12031.
108. Voorn-Brouwer T, Kragt A, Tabak HF et al. Peroxisomal membrane proteins are properly targeted to peroxisomes in the absence of COPI- and COPII-mediated vesicular ransport. J Cell Sci 2001; 114:2199-2204.
109. Titorenko VI, Rachubinski RA. Mutants of the yeast Yarrowia lipolytica defective in protein exit from the endoplasmic reticulum are also defective in peroxisome biogenesis. Mol Cell Biol 1998; 18:2789-2803.
110. Geuze HJ, Murk JL, Stroobants AK et al. Involvement of the endoplasmic reticulum in peroxisome formation. Mol Biol Cell 2003; 14:2900-2907.
111. De Duve C. Evolution of the peroxisome. Ann NY Acad Sci 1969; 168:369-381.

Index

N

O

P

Q

R

S

T

U

Y

Z